DIMENSION THEORY

PRINCETON MATHEMATICAL SERIES

Editors: MARSTON MORSE and A. W. TUCKER

1. The Classical Groups, Their Invariants and Representation. By HERMANN WEYL.
3. An Introduction to Differential Geometry with Use of the Tensor Calculus. By LUTHER PFAHLER EISENHART.
4. Dimension Theory. By WITOLD HUREWICZ and HENRY WALLMAN.
5. The Analytic Foundations of Celestial Mechanics. By AUREL WINTNER.
6. The Laplace Transform. By DAVID VERNON WIDDER.
7. Integration. By EDWARD JAMES McSHANE.
8. Theory of Lie Groups: I. By CLAUDE CHEVALLEY.
9. Mathematical Methods of Statistics. By HARALD CRAMÉR.
10. Several Complex Variables. By SALOMON BOCHNER and WILLIAM TED MARTIN.
11. Introduction to Topology. By SOLOMON LEFSCHETZ.
12. Algebraic Geometry and Topology. Edited by R. H. FOX, D. C. SPENCER, and A. W. TUCKER.
14. The Topology of Fibre Bundles. By NORMAN STEENROD.
15. Foundations of Algebraic Topology. By SAMUEL EILENBERG and NORMAN STEENROD.
16. Functionals of Finite Riemann Surfaces. By MENAHEM SCHIFFER and DONALD C. SPENCER.
17. Introduction to Mathematical Logic, Vol. I. By ALONZO CHURCH.
18. Algebraic Geometry. By SOLOMON LEFSCHETZ.
19. Homological Algebra. By HENRI CARTAN and SAMUEL EILENBERG.
20. The Convolution Transform. By I. I. HIRSCHMAN and D. V. WIDDER.
21. Geometric Integration Theory. By HASSLER WHITNEY.
22. Qualitative Theory of Differential Equations. By V. V. NEMICKII and V. V. STEPANOV. Translated under the direction of SOLOMON LEFSCHETZ.
23. Topological Analysis, revised edition. By GORDON T. WHYBURN.
24. Analytic Functions. By NEVANLINNA, BEHNKE and GRAUERT, et al.
25. Continuous Geometry. By JOHN VON NEUMANN. Foreword by ISRAEL HALPERIN.
26. Riemann Surfaces. By L. AHLFORS and L. SARIO.
27. Differential and Combinatorial Topology. Edited by S. S. CAIRNS.

DIMENSION THEORY

By

WITOLD HUREWICZ

AND

HENRY WALLMAN

PRINCETON

PRINCETON UNIVERSITY PRESS

LONDON: OXFORD UNIVERSITY PRESS

1948

Revised edition 1948

Eighth Printing, 1968

Printed in the United States of America

Preface

In this book it has been the aim of the authors to give a connected and simple account of the most essential parts of dimension theory. Only those topics were chosen which are of interest to the general worker in mathematics as well as the specialist in topology.

Since the appearance of Karl Menger's well-known "Dimensionstheorie" in 1928, there have occurred important advances in the theory, both in content and method. These advances justify a new treatment, and in the present book great emphasis has been laid on the modern techniques of function spaces and mappings in spheres.

The algebraically minded reader will find in Chapter VIII a concise exposition of modern homology theory, with applications to dimension.

Historical references are made solely for the guidance of the beginning student, and no attempt has been made to attain completeness in this respect.

The authors wish to express their thanks to Drs. Samuel Eilenberg and William Flexner, who gave stimulating advice in preparing the manuscript, and to Mr. James Dugundji, who carefully read the proof and made valuable suggestions.

WITOLD HUREWICZ
HENRY WALLMAN

Chapel Hill, North Carolina
Madison, Wisconsin
October 1941

Contents

DIMENSION THEORY

CHAPTER I

Introduction

1. The modern concept of dimension

"Of all the theorems of analysis situs, the most important is that which we express by saying that space has three dimensions. It is this proposition that we are about to consider, and we shall put the question in these terms: when we say that space has three dimensions, what do we mean? . . .

" . . . if to *divide* a continuum it suffices to consider as cuts a certain number of elements all distinguishable from one another, we say that this continuum is of *one dimension*; if, on the contrary, to divide a continuum it is necessary to consider as cuts a system of elements themselves forming one or several continua, we shall say that this continuum is of *several dimensions.*

"If to divide a continuum C, cuts which form one or several continua of one dimension suffice, we shall say that C is a continuum of *two dimensions*; if cuts which form one or several continua of at most two dimensions suffice, we shall say that C is a continuum of *three dimensions*; and so on.

"To justify this definition it is necessary to see whether it is in this way that geometers introduce the notion of three dimensions at the beginning of their works. Now, what do we see? Usually they begin by defining surfaces as the boundaries of solids or pieces of space, lines as the boundaries of surfaces, points as the boundaries of lines, and they state that the same procedure can not be carried further.

"This is just the idea given above: to divide space, cuts that are called surfaces are necessary; to divide surfaces, cuts that are called lines are necessary; to divide lines, cuts that are called points are necessary; we can go no further and a point can not be divided, a point not being a continuum. Then lines, which can be divided by cuts which are not continua, will be continua of one dimension; surfaces, which can be divided by continuous cuts of one dimension, will be continua of two dimensions; and finally space, which can be divided by continuous cuts of two dimensions, will be a continuum of three dimensions."

These words were written by Poincaré in 1912, in the last year of his life. Writing in a philosophical journal,* Poincaré was concerned only with putting forth an intuitive concept of dimension and not an exact mathematical formulation. Poincaré had, however, penetrated very deep, in stressing the inductive nature of the geometric meaning of dimension and the possibility of disconnecting a space by subsets of lower dimension. One year later Brouwer† constructed on Poincaré's foundation a precise and topologically invariant definition of dimension, which for a very wide class of spaces (locally-connected separable metric) is equivalent to the one we use today.

Brouwer's paper remained almost unnoticed for several years. Then in 1922, independently of Brouwer, and of each other, Menger and Urysohn recreated Brouwer's concept, with important improvements; and what is more noteworthy, they justified the new concept by making it the cornerstone of an extremely beautiful and fruitful theory which brought unity and order to a large domain of geometry.

The definition of dimension we shall adopt in this book (see page 24) is due to Menger and Urysohn. In the formulation of Menger, it reads:

a) the empty set has dimension -1,

b) the dimension of a space is the least integer n for which every point has arbitrarily small neighborhoods whose boundaries have dimension less than n.

It is the opinion of the authors that none of the several other possible definitions of dimension has the immediate intuitive appeal of this one and none leads so elegantly to the existing theory.

2. Previous concepts of dimension

Before the advent of set theory mathematicians used dimension in only a vague sense. A configuration was said to be n-dimensional if the least number of real parameters needed to describe its points, in some unspecified way, was n. The dangers and inconsistencies in this approach were brought into clear view by two celebrated discoveries of the last part of the 19th century: Cantor's 1:1 correspondence between the points of a line and the points of a plane, and Peano's continuous mapping of an interval on the whole of a square. The first exploded the feeling that a plane is richer in points than a line, and showed that dimension can be changed by a 1:1 transformation. The second contradicted the belief that the dimension of a space could be defined as

* *Revue de métaphysique et de morale*, p. 486.

† Über den natürlichen Dimensionsbegriff, *Journ. f. Math.* 142 (1913), pp. 146–152.

the least number of continuous real parameters required to describe the space, and showed that dimension can be raised by a one-valued continuous transformation.

An extremely important question was left open (and not answered until 1911, by Brouwer): Is it possible to establish a correspondence between Euclidean n-space (the ordinary space of n real variables) and Euclidean m-space combining the features of both Cantor's and Peano's constructions, i.e. a correspondence which is both 1:1 *and* continuous? The question is crucial since the existence of a transformation of the stated type between Euclidean n-space and Euclidean m-space would signify that dimension (in the natural sense that Euclidean n-space has dimension n) has no topological meaning whatsoever! The class of topological transformations would in consequence be much too wide to be of any real geometric use.

3. Topological invariance of the dimension of Euclidean spaces

The first proof that Euclidean n-space and Euclidean m-space are not homeomorphic unless n equals m was given by Brouwer in his famous paper: Beweis der Invarianz der Dimensionenzahl, *Math. Ann.* 70 (1911), pp. 161–165. However, this proof did not explicitly reveal any simple topological property of Euclidean n-space distinguishing it from Euclidean m-space and responsible for the non-existence of a homeomorphism between the two. More penetrating, therefore, was Brouwer's procedure in 1913* when he introduced his "Dimensionsgrad," an integer-valued function of a space which was topologically invariant by its very definition. Brouwer showed that the "Dimensionsgrad" of Euclidean n-space is precisely n (and therefore deserved its name).

Meanwhile Lebesgue had approached in another way the proof that the dimension of a Euclidean space is topologically invariant. He had observed† that a square can be covered by arbitrarily small "bricks" in such a way that no point of the square is contained in more than three of these bricks; but that if the bricks are sufficiently small, at least three have a point in common. In a similar way a cube in Euclidean n-space can be decomposed into arbitrarily small bricks so that not more than $n + 1$ of these bricks meet. Lebesgue conjectured that this number $n + 1$ could not be reduced further, i.e. for any decomposition in sufficiently small bricks there must be a point common

* Über den natürlichen Dimensionsbegriff, loc. cit.

† "Sur la non applicabilité de deux domaines appartenant à des espaces de n et $n + p$ dimensions," *Math. Ann.* 70 (1911), pp. 166–168.

to at least $n + 1$ of the bricks. (The first proof of this theorem was given by Brouwer.*) Lebesgue's theorem also displays a topological property of Euclidean n-space distinguishing it from Euclidean m-space and therefore it also implies the topological invariance of the dimension of Euclidean spaces.

4. Dimension of general sets

The new concept of dimension, as we have already seen, gave a precise meaning to the statement that Euclidean n-space has dimension n, and thereby clarified considerably the entire structure of topology. Another feature which made the new dimension concept a milestone in geometry was the generality of the objects to which it could be applied. The lack of a precise definition of dimension, however unsatisfactory from an esthetic and methodological point of view, caused no real difficulty so long as geometry was confined to the study of relatively simple figures, such as polyhedra and manifolds. No doubt could arise, in each particular case, as to what dimension to assign to each of these figures. This situation changed radically, following the discoveries of Cantor, with the development of point-set theory. This new branch of mathematics tremendously enlarged the domain of what could be considered as "geometrical objects" and revealed configurations of complexity never before dreamt of. To associate with each of these objects a number which might reasonably be called a dimension was by no means a trivial task. What, for instance, was one to take as the dimension of the indecomposable continuum of Brouwer, or of Sierpiński's "curve" each of whose points is a ramification point?

Dimension-theory gives a complete answer to these questions. It assigns to every set of points in a Euclidean space (and even to every subset of Hilbert space†), no matter how "pathological," an integer which on intuitive and formal grounds strongly deserves to be called its dimension.

5. Different approaches to dimension

Before we proceed to a systematic study of dimension-theory let us pause to consider other possible ways of defining dimension.

We have already mentioned Lebesgue's method of proving the invariance theorem for the dimension of Euclidean spaces. His procedure can be used very well to establish a general concept of dimension: the Lebesgue dimension of a space is the least integer n with the

* Über den natürlichen Dimensionsbegriff, loc. cit.

† See Appendix for remarks concerning more general spaces.

property that the space may be decomposed into arbitrarily small domains not more than $n + 1$ of which meet. It turns out (see Chapter V) that this method of introducing dimension coincides with that due to Brouwer, Menger, and Urysohn. We shall now give other examples showing how topological investigations of very different nature all lead to the same concept of dimension.

A)* Let

$$f_i(x_1, \cdots, x_n), \qquad i = 1, \cdots, m \tag{1}$$

be m continuous real valued functions of n real unknowns, or what is the same, m continuous real-valued functions of a point in Euclidean n-space. It is one of the basic facts of analysis that the system of m equations in n unknowns,

$$f_i(x_1, \cdots, x_n) = 0, \tag{2}$$

has, in general, no solution if $m > n$. The words "in general" may be made precise as follows: by modifying the functions f_i very little one can obtain new continuous functions g_i such that the new system

$$g_i(x_1, \cdots, x_n) = 0 \tag{3}$$

has no solution. On the other hand, there do exist sets of n equations in n unknowns which are solvable, and which remain solvable after any sufficiently small modification of their left members. This property of Euclidean n-space can be made the basis of a general concept of dimension. A space X would be called n-dimensional if n is the largest integer for which there exist n continuous real-valued functions (1) defined over X such that the system of equations (2) has a solution which is essential in the sense explained above. It turns out that this "dimension" is again the same as the dimension of Brouwer, Menger, and Urysohn (see Chapter VI).

B) A modification of A) is this problem. Consider continuous transformations of a space in an n-sphere. Every point of the n-sphere may be regarded as a vector of length unity (a "direction") in Euclidean $(n + 1)$-space, so that instead of continuous transformations in the n-sphere, one may speak of continuous fields of non-vanishing $(n + 1)$-vectors. Suppose C is a closed set in a space X. Given a continuous field of non-vanishing $(n + 1)$-vectors on C, is it possible to extend this field, without changing the vectors on C, to a continuous field of non-

* The questions discussed in A), B), and C) are closely related to each other, and were raised by Alexandroff in a paper whose results underlie much of Chapters VI and VIII: Dimensienstheorie, *Math. Ann.* 106 (1932), pp. 161–238.

vanishing $(n + 1)$-vectors defined over all of X? It turns out that the dimension of X is the least integer n for which such an extension is possible for each closed set C and each continuous field of non-vanishing $(n + 1)$-vectors on C; in terms of mappings in the n-sphere, the least integer n with the property that any continuous mapping of any closed subset C into the n-sphere can be extended to all of X (see Chapter VI).

C) Another approach to dimension arises from homology theory. Consider 1-cycles (roughly, continuous closed curves) in a 2-dimensional manifold. Some of these bound a 2-dimensional part of the manifold, or in the notation of homology theory, are bounding cycles. On the other hand no 2-cycle (with the obvious exception of the vacuous 2-cycle all of whose coefficients are zero) can bound in the 2-manifold, because there is nothing 3-dimensional for it to bound. In a similar way an n-dimensional manifold contains non-vacuous bounding m-cycles for every m less than n, but contains only vacuous bounding n-cycles. Now, homology theory can be applied to arbitrary compact metric spaces. One may then define the "homology-dimension" of a compact metric space as the largest integer n for which there exist, with suitably chosen coefficients, non-vacuous bounding $(n - 1)$-cycles. The homology-dimension so defined turns out to be the same as our standard dimension (see Chapter VIII).

D) The intuitive perception of dimension associates with the word 1-dimensional, objects having length (or linear measure), with the word 2-dimensional, objects having area (or 2-dimensional measure), with the word 3-dimensional, objects having volume (or 3-dimensional measure), and so on. An attempt to make this intuitive feeling precise meets the obstacle that dimension is a topological concept while measure is a metrical concept. However, let us consider with a given metric space X all the metrics compatible with its topological structure. We find (see Chapter VII) that the dimension of X can be characterized as the largest real number p for which X, *in each metrization*, has positive p-dimensional Hausdorff measure.

6. Remarks

In this book we assume only a very elementary knowledge of point-set topology, such as is contained, for example, in the first chapters of Alexandroff-Hopf's *Topologie*, Julius Springer, Berlin, 1935; Kuratowski's *Topologie* I, Monografje Matematyczne, Warsaw, 1933;

Menger's *Dimensionstheorie*, B. G. Teubner, Leipzig, 1928. It will, however, be possible for the reader to refresh his acquaintance with the fundamentals of topology by the use of the index. The index contains, besides references to the definitions and results of dimension-theory, a considerable number of very brief discussions (and even some proofs) of topics in general topology which the development requires.

A large number of illustrative examples are included in the book, many of them without proof; these should be regarded as exercises.

Mathematical assertions of subsidiary importance are called Propositions. References to these are made according to the following scheme:

"By Proposition A)" means by Proposition A) of the same section and chapter in which the reference occurs.

"By Proposition 2 A)" means by Proposition A) of Section 2 of the same chapter in which the reference occurs.

"By Proposition III 2 A)" means by Proposition A) of Section 2 of Chapter III.

Throughout this book ***all spaces are separable metric,*** unless the contrary is explicitly stated. This limitation is made because there arise grave difficulties in extending dimension-theory to more general spaces. A brief discussion of some of these difficulties is given in the Appendix.

CHAPTER II

Dimension 0

Topology consists essentially in the study of the connectivity structure of spaces. The concept of a connected space, which in its present form is due to Hausdorff and Lennes, may be considered the root-concept from which is derived, directly or indirectly, the bulk of the important concepts of topology (homology or "algebraic connectivity" theory, local connectedness, dimension, etc.).

A space is *connected* if it cannot be split into two non-empty disjoint open sets. Equivalently: a space is connected if, except for the empty set and the whole space, there are no sets whose boundaries* are empty.

In this chapter we are concerned with spaces which are disconnected in an exceedingly strong sense, viz. have so many open sets whose boundaries are empty that every point may be enclosed in arbitrarily small sets of this type.

1. Definition of dimension 0

Definition II 1. A space X has *dimension* 0 *at a point* p if p has arbitrarily small neighborhoods† with empty boundaries, i.e. if for each neighborhood U of p there exists a neighborhood V of p such that

$$V \subset U,$$

$$\text{bdry } V = 0.$$

A non-empty space X has *dimension* 0, dim $X = 0$, if X has dimension 0 at each of its points.

A) It is obvious that the property of being 0-dimensional, or of being 0-dimensional at a point p, is a topological invariant.

B) A 0-dimensional space can also be defined as a non-empty space in which there is a basis* for the open sets made up of sets which are at the same time open and closed.

Example II 1. Every non-empty finite or countable space X is

* See index. Observe that any set whose boundary is empty is both open and closed, and conversely.

† By a neighborhood of a point we mean any open set containing the point.

0-dimensional.* For given any neighborhood U of any point p let ρ be a positive real number such that the spherical neighborhood of radius ρ about p (the set of all points whose distance from p is less than ρ) is contained in U. Let $x_1, x_2, \cdots$ be an enumeration of X and $d(x_i, p)$ the distance from x_i to p. There exists a positive real number ρ' which is less than ρ and different from all the $d(x_i, p)$. The spherical neighborhood of radius ρ' about p is then contained in U and its boundary is empty. Hence X is 0-dimensional.

In particular the set $\mathcal{R}$ of rational real numbers is 0-dimensional.

EXAMPLE II 2. The set $\mathcal{J}$ of irrational real numbers is 0-dimensional. For given any neighborhood U of an irrational point p there exist rational numbers ρ and σ such that $\rho < p < \sigma$ and the set V of irrational numbers between ρ and σ is contained in U. In the space $\mathcal{J}$ of irrationals V is open, and has an empty boundary because every irrational point which is a cluster-point† of V is between ρ and σ and hence belongs to V.

EXAMPLE II 3. The Cantor discontinuum‡ $\mathcal{C}$ (the set of all real numbers expressible in the form $\sum_1^\infty a_n/3^n$ where $a_n = 0$ or 2) is 0-dimensional.

EXAMPLE II 4. Any set of real numbers containing no interval is 0-dimensional.

EXAMPLE II 5. The set $\mathcal{J}_2$ of points in the plane both of whose coordinates are irrational is 0-dimensional. For any such point is contained in arbitrarily small rectangles bounded by lines having rational intercepts with the coordinate axes and intersecting them at right angles, and the boundaries of such rectangles do not meet $\mathcal{J}_2$.

EXAMPLE II 6. The set $\mathcal{R}_2^1$ of points in the plane exactly one of whose coordinates is rational is 0-dimensional. For any such point is contained in arbitrarily small rectangles bounded by lines having rational intercepts with the coordinate axes and intersecting them at 45°, and the boundaries of such rectangles do not meet $\mathcal{R}_2^1$.

EXAMPLE II 7. The set $\mathcal{R}_n$ of points in Euclidean n-space§ E_n all of whose coordinates are rational is 0-dimensional. For $\mathcal{R}_n$ is countable.

EXAMPLE II 8. The set $\mathcal{J}_n$ of points in E_n all of whose coordinates are irrational is 0-dimensional. This is a simple generalization of Example II 5.

REMARK. Suppose $0 \leqq m \leqq n$. Denote by $\mathcal{R}_n^m$ the set of points in

* Do not forget that unless the contrary is explicitly stated all spaces considered in this book are separable metric.

† See index.

‡ See Hausdorff: *Mengenlehre,* de Gruyter, Berlin 2nd ed. 1927, p. 134.

§ See index.

E_n exactly m of whose coordinates are rational. In Examples II 7 and II 8 we have seen that $\mathcal{R}_n^n = \mathcal{R}_n$ and $\mathcal{R}_n^0 = \mathcal{J}_n$ are 0-dimensional. It is true (Example II 12) that $\mathcal{R}_n^m$ is 0-dimensional for each m and n, but the proof depends essentially on the "Sum Theorem for 0-dimensional Sets," Theorem II 2, and the simple proof of Example II 6 cannot be generalized.

EXAMPLE II 9. The set $\mathcal{R}_\omega'$ of points in the Hilbert* cube I_ω all of whose coordinates are rational is 0-dimensional. (This set is not countable.)

Suppose

$$a = (a_1, a_2, \cdots)$$

is an arbitrary point in I_ω and U is a neighborhood of a in I_ω. By taking n large enough and p_i, q_i sufficiently close to a_i, $p_i < a_i < q_i$, $i = 1, \cdots, n$, one gets a neighborhood of a contained in U consisting of the points

$$x = (x_1, x_2, \cdots)$$

of I_ω whose first n coordinates are restricted by

$$p_i < x_i < q_i \tag{1}$$

(and whose other coordinates are restricted only by

$$|x_i| \leqq 1/i, \tag{2}$$

((2) being, of course, always present in I_ω).†

Now, suppose $a \in \mathcal{R}_\omega'$. By taking p_i and q_i irrational we get a neighborhood V of a each of whose boundary points in I_ω has at least one

* See index.

† We have to show that for each $\epsilon > 0$ we can find an integer n and a positive real number δ such that if $q_i - p_i < \delta$ for $i \leqq n$ then all x satisfying (1) and (2) also satisfy

$$\left[\sum_{i=1}^{\infty} (x_i - a_i)^2\right]^{\frac{1}{2}} < \epsilon. \tag{3}$$

To do this choose n so that

$$\sum_{i=n}^{\infty} \frac{1}{i^2} < \frac{1}{8}\epsilon^2$$

and δ so that

$$n\delta^2 < \tfrac{1}{2}\epsilon^2.$$

If $q_i - p_i$ is less than δ for $i \leqq n$, we have for all x satisfying (1) and (2) that

$$\sum_{i=1}^{\infty} (x_i - a_i)^2 < n\delta^2 + \sum_{i=n}^{\infty} \left(\frac{2}{i}\right)^2 < \tfrac{1}{2}\epsilon^2 + \tfrac{1}{2}\epsilon^2 = \epsilon^2;$$

which proves (3).

irrational coordinate. Hence V has an empty boundary in $\mathcal{R}_\omega'$, and this proves that $\mathcal{R}_\omega'$ is 0-dimensional.

Example II 10. The set $\mathcal{I}_\omega'$ of points in the Hilbert cube all of whose coordinates are irrational is 0-dimensional. The proof parallels that of Example II 9.

Example II 11. The set $\mathcal{R}_\omega$ of points in Hilbert space all of whose coordinates are rational is *not* 0-dimensional.* (Contrast this statement with that of Example II 9.) It suffices to show that any bounded neighborhood U in $\mathcal{R}_\omega$ of the origin has a non-empty boundary. Consider the straight line $-\infty < x_1 < \infty, 0 = x_2 = x_3 = \cdots$. This contains points in U and points in $\mathcal{R}_\omega - U$. It follows that one can find a rational number a_1 such that the point

$$p^1 = (a_1, 0, 0, \cdots)$$

is contained in U and has distance less than 1 from $\mathcal{R}_\omega - U$. Similarly, by considering the straight line $x_1 = a_1, -\infty < x_2 < \infty, 0 = x_3 = x_4 = \cdots$, we determine a point

$$p^2 = (a_1, a_2, 0, \cdots),$$

with rational a_2, and such that p^2 is in U and has distance less than $\frac{1}{2}$ from $\mathcal{R}_\omega - U$. By induction we determine a sequence $\{p^n\}$:

$$p^n = (a_1, a_2, \cdots, a_n, 0, 0, \cdots),$$

where each a_n is rational, each p^n is in U, and

$$d(p^n, \mathcal{R}_\omega - U) < \frac{1}{n} \cdot$$

Then it easily follows that the point p:†

$$p = (a_1, a_2, \cdots, a_k, a_{k+1}, \cdots),$$

whose k^{th} coordinate is a_k, is a boundary point of U.

Theorem II 1. *A non-empty subset of a* 0*-dimensional space is* 0*-dimensional.*

Proof. Let p be any point in the non-empty subset X' of the 0-dimensional space X and let U' be any neighborhood of p in X'. Then there exists a neighborhood U in X of p such that

* This proof is due to Paul Erdös, The dimension of rational points in Hilbert space, *Ann. Math.*, 41 (1940), 734–736. In fact, dim $\mathcal{R}_\omega = 1$ (see Example III 5).

† p is actually a point of $\mathcal{R}_\omega$, for $\sum_1^n a_i^2 < d^2$, $n = 1, 2, \cdots$, where d is the diameter of U; hence $\sum_1^\infty a_i^2 < \infty$.

$$U' = UX'.$$

Since X is 0-dimensional there exists a set V which is both open and closed in X such that

$$p \varepsilon V \subset U.$$

Let

$$V' = VX'.$$

Then V' is both open and closed in X' and

$$p \varepsilon V' \subset U',$$

so that X' is 0-dimensional.

2. Separation of subsets

Definition II 2. If A_1, A_2 and B are mutually disjoint subsets of a space X, we say that A_1 and A_2 are *separated in* X *by* B if $X - B$ can be split into two disjoint sets, open in $X - B$ and containing A_1 and A_2 respectively, i.e. if there exist A_1' and A_2' for which

$$X - B = A_1' + A_2',$$
$$A_1 \subset A_1', \qquad A_2 \subset A_2',$$
$$A_1' A_2' = 0,$$

with A_1' and A_2' both open in $X - B$ (or what is the same, both closed in $X - B$).

If A_1 and A_2 are separated by the empty set we say simply that A_1 and A_2 are *separated in* X.

A) A_1 and A_2 are separated if and only if there exists a set A_1' such that

$$A_1 \subset A_1'$$
$$A_1' \cdot A_2 = 0$$

and A_1' is both open and closed, i.e. has empty boundary. For A_2' is then $X - A_1'$.

B) We now prove that Definition II 1 is equivalent to

Definition II 1′. A non-empty space has dimension 0 if every point p and every closed set C not containing p can be separated.*

* Throughout this book we shall make no distinction in notation between the point p and the set consisting of the single point p.

PROOF. Suppose X is 0-dimensional in the sense of Definition II 1. Then since $X - C$ is a neighborhood of p there exists a set V such that

$$p \varepsilon V \subset X - C$$

with V both open and closed. From

$$V \cdot C = 0$$

and Proposition A) it follows that p and C are separated, as required by Definition II 1′. The converse is proved similarly.

C) A connected 0-dimensional space consists of only one point.

PROOF. Suppose a 0-dimensional space X contains two distinct points p and q. Definition II 1′ shows that p and q are separated. Hence X is disconnected.

D) A 0-dimensional space is totally-disconnected, i.e. no connected subset contains more than one point.

PROOF. This follows from Theorem II 1 and Proposition C).

E) It is obvious from Definition II 1′ that a space is 0-dimensional if any two disjoint closed sets in it can be separated. We now prove conversely that* if a space is 0-dimensional, any two disjoint closed sets in it can be separated.

PROOF. Let X be 0-dimensional. We know from Definition II 1′ that any point p in X can be separated from any closed set not containing p. Let C and K be two disjoint closed sets in X. We have to demonstrate a separation of C and K in X.

For each point p of X either $p \cdot C = 0$ or $p \cdot K = 0$. Hence there exist neighborhoods $U(p)$ for each p which are both open and closed and such that either $U(p) \cdot C = 0$ or $U(p) \cdot K = 0$. Since X has a countable open basis there exists a sequence $U_1, U_2, \cdots$ of these $U(p)$ whose sum is X (Alexandroff-Hopf: *Topologie*, Julius Springer, Berlin 1935, p. 78; this book will hereafter be cited as AH). We define a new sequence of sets V_i as follows:

$$V_1 = U_1,$$

* The existence of a countable basis for the open sets of X is essential, as is shown in the Appendix; see assertion (b) in the Appendix.

$$V_i = U_i - \sum_{k=1}^{i-1} U_k = U_i \left(X - \sum_{k=1}^{i-1} U_k \right), \qquad i = 2, 3, \cdots.$$

Then we have

(1) $$X = \sum_1^{\infty} V_i,$$

(2) $$V_i V_j = 0 \quad \text{if} \quad i \neq j,$$

(3) $$V_i \quad \text{is open,}$$

(4) $$\text{either} \quad V_i C = 0 \quad \text{or} \quad V_i K = 0.$$

(1), (2), and (4) are obvious. To prove (3) note that

$$\sum_{k=1}^{i-1} U_k$$

is closed, so that

$$X - \sum_{k=1}^{i-1} U_k$$

is open; hence V_i, as the intersection of this open set with the open set U_i, is itself open.

Let C' be the sum of all V_i for which $V_i \cdot K = 0$, and K' the sum of the remaining V_i. Then

$$X = C' + K' \qquad \text{by (1),}$$

$$C'K' = 0 \qquad \text{by (2),}$$

$$C' \text{ and } K' \text{ are open} \qquad \text{by (3), and}$$

$$C'K + K'C = 0 \qquad \text{by (4).}$$

It follows that $C \subset C'$ and $K \subset K'$. The desired separation is thus given by C' and K'.

F) If C_1 and C_2 are disjoint closed sets in a space X and A is a 0-dimensional subset of X, then there exists a closed set B separating C_1 and C_2 such that $AB = 0$.

PROOF. Since X is normal* there exist open sets U_1 and U_2 for which

* A space S is *normal* (AH, p. 68) if for any two disjoint closed subsets C_1 and C_2 there exist open subsets V_1 and V_2 such that

$$C_1 \subset V_1, \qquad C_2 \subset V_2, \quad \text{and}$$
$$V_1 V_2 = 0.$$

$$(5) \qquad C_1 \subset U_1, \qquad C_2 \subset U_2, \quad \text{and}$$
$$\overline{U}_1\overline{U}_2 = 0.$$

The disjoint sets $\overline{U}_1A$ and $\overline{U}_2A$ are closed in A and can therefore be separated in A, by Proposition E), since $\dim A = 0$. We then have disjoint sets C_1' and C_2' satisfying

$$A = C_1' + C_2'$$
$$\overline{U}_1A \subset C_1', \qquad \overline{U}_2A \subset C_2',$$

with C_1' and C_2' both open and closed in A. Hence

$$(6) \qquad C_1'\overline{U}_2 + C_2'\overline{U}_1 = 0,$$

$$(7) \qquad C_1'\overline{C}_2' + \overline{C}_1'C_2' = 0.$$

(5) and (6) imply

$$(8) \qquad C_1'\overline{C}_2 + C_2'\overline{C}_1 = 0.$$

Further, since U_1 and U_2 are open, (6) implies $\overline{C}_1'U_2 + \overline{C}_2'U_1 = 0$, and hence by (5),

$$(9) \qquad \overline{C}_1'C_2 + \overline{C}_2'C_1 = 0.$$

From (7), (8), and (9) together with $\overline{C}_1\overline{C}_2 = C_1C_2 = 0$ it follows that neither of the disjoint sets $C_1 + C_1'$ and $C_2 + C_2'$ contains a cluster point of the other. The complete normality* of X insures the existence of an open set W such that

$$C_1 + C_1' \subset W, \quad \text{and}$$
$$\overline{W}(C_2 + C_2') = 0.$$

It is a simple consequence of normality that there exist open sets U_1 and U_2 such that

$$C_1 \subset U_1, \qquad C_2 \subset U_2, \quad \text{and}$$
$$\overline{U}_1\overline{U}_2 = 0.$$

Every metric space is normal (AH, p. 68).

* A space X is *completely normal* (AH, p. 69) if every subspace of X is normal. It can be shown (Urysohn: Über die Mächtigkeit der zusammenhängenden Mengen, *Math. Ann.* 94 (1925), pp. 262–295, in particular p. 284) that for any two disjoint subsets X_1 and X_2 of a completely normal space neither containing a cluster point of the other, there exist open sets W_1 and W_2 such that

$$X_1 \subset W_1, \qquad X_2 \subset W_2, \quad \text{and}$$
$$W_1W_2 = 0.$$

It is clear that $\overline{W}_1 \cdot X_2 = 0$.

Every metric space is completely normal, because every subspace of a metric space is metric.

The boundary $B = \overline{W} - W$ separates C_1 and C_2 and is disjoint from $C_1' + C_2' = A$. This proves F).

3. The sum-theorem for 0-dimensional sets

The sum of zero-dimensional sets need not be zero-dimensional, as we see from the decomposition of the line into the rational numbers and the irrational numbers, or into its distinct points. However

Theorem II 2. The Sum Theorem for 0-dimensional Sets. *A space which is the* **countable** *sum of* 0-*dimensional* **closed** *subsets is itself* 0-*dimensional.*

PROOF. Suppose

$$X = C_1 + C_2 + \cdots + C_i + \cdots,$$

each C_i closed and 0-dimensional. Let K and L be two disjoint closed sets in X. We shall show that K and L can be separated.

KC_1 and LC_1 are disjoint closed subsets of the 0-dimensional space C_1. Hence (2E)) there exist subsets A_1 and B_1 of C_1 which are closed in C_1, and therefore in X, such that

$$KC_1 \subset A_1, \qquad LC_1 \subset B_1,$$
$$A_1 + B_1 = C_1, \qquad A_1B_1 = 0.$$

The sets $K + A_1$ and $L + B_1$ are closed and disjoint in X. By the normality of X we have open sets G_1 and H_1 for which

$$K + A_1 \subset G_1, \qquad L + B_1 \subset H_1,$$
$$\overline{G}_1\overline{H}_1 = 0.$$

Therefore

$$G_1 + H_1 \supset C_1$$
$$K \subset G_1, \qquad L \subset H_1,$$
$$\overline{G}_1\overline{H}_1 = 0.$$

Now repeat this process, replacing K and L by $\overline{G}_1$ and $\overline{H}_1$ and C_1 by C_2. This yields open sets G_2 and H_2 for which

$$G_2 + H_2 \supset C_2,$$
$$\overline{G}_1 \subset G_2, \qquad \overline{H}_1 \subset H_2,$$
$$\overline{G}_2\overline{H}_2 = 0.$$

By induction we construct sequences $\{G_i\}$ and $\{H_i\}$ of sets open in X for which

$$G_i + H_i \supset C_i$$
$$\overline{G}_{i-1} \subset G_i, \qquad \overline{H}_{i-1} \subset H_i,$$
$$\overline{G}_i \overline{H}_i = 0.$$

Let

$$G = \sum G_i, \qquad H = \sum H_i.$$

Then G and H are disjoint open sets,

$$G + H \supset \sum C_i = X$$

and

$$K \subset G, \qquad L \subset H;$$

this is the desired separation.

Corollary 1. *A space which is the countable sum of 0-dimensional F_σ sets* is 0-dimensional.*

Corollary 2. *The sum of two 0-dimensional subsets of a space X at least one of which is closed is 0-dimensional.*

Proof. Let A and B be 0-dimensional and B closed. Then $(A + B) - B$ is open in $A + B$. As an open set in a metric space it is an F_σ in $A + B$. Corollary 2 then follows from Corollary 1 and the equation

$$A + B = B + [(A + B) - B].$$

Corollary 3. *A 0-dimensional space remains 0-dimensional after the adjunction of a single point.*†

Proof. Obvious from Corollary 2.

Example II 12. Suppose $0 \leqq m \leqq n$. Denote by $\mathfrak{R}_n^m$ the set of points in Euclidean n-space E_n exactly m of whose coordinates are rational. Then $\mathfrak{R}_n^m$ is 0-dimensional.

For each selection of m indices $i_1, \cdots, i_m$ out of the range $1, \cdots, n$, and each selection of m rational numbers $r_1, \cdots, r_m$ we have an $(n - m)$-dimensional linear subspace of E_n determined by the equations

$$(1) \qquad x_{i_1} = r_1, x_{i_2} = r_2, \cdots, x_{i_m} = r_m.$$

The subset of (1) made up of the points none of whose remaining co-

* By an F_σ in a space we mean any countable sum of closed subsets; see Kuratowski: *Topologie* I, Monografje Matematyczne, Warsaw 1933, p. 21. In a metric space any open set is an F_σ (Kuratowski *Topologie* I, loc. cit., p. 78).

† Assuming, of course, that the enlarged space is also separable metric.

ordinates is rational we denote by C_i. Each C_i is congruent to $\mathcal{J}_{n-m}$ and is therefore 0-dimensional (Example II 8). It is clear that each C_i is closed in $\mathcal{R}_n^m$ and that the sum of the C_i just fills out $\mathcal{R}_n^m$. Since the collection of the C_i is countable, the 0-dimensional sum theorem implies that dim $\mathcal{R}_n^m = 0$.

EXAMPLE II 13. Suppose $0 \leqq m$. Denote by $\mathcal{R}_\omega^m$ the set of points in the Hilbert-cube exactly m of whose coordinates are rational. Then $\mathcal{R}_\omega^m$ is 0-dimensional. (The proof is like that of Example II 12 and makes use of Example II 10.)

4. Compact spaces

Consider the following four properties of a space X:

(0) X is totally disconnected.

(1) Any two distinct points in X can be separated.

(2) Any point can be separated from any closed set not containing it, i.e. X is 0-dimensional (see Definition II 1′).

(3) Any two disjoint closed sets can be separated.

Obviously (3) implies (2), (2) implies (1), and (1) implies (0). Conversely (see Proposition 2E)), (2) implies (3) (for the spaces with countable basis with which this book is almost entirely concerned; for spaces without countable basis (2) does not imply (3): see the Appendix, p. 155). Properties (0), (1), and (2), however are *not* equivalent, even for separable metric spaces:

EXAMPLE II 14. Sierpiński: Sur les ensembles connexes et non connexes, *Fund. Math.* 2 (1921), pp. 81–95, gives on p. 88 an example of a subset of the plane satisfying (0) but not (1).

EXAMPLE II 15. We have learned in Example II 11 that $\mathcal{R}_\omega$ does not satisfy (2). On the other hand, it does satisfy (1). For, let p and q be two points of $\mathcal{R}_\omega$ and let i be an index such that the ith coordinate p_i of p differs from the ith coordinate q_i of q; p_i and q_i are of course rational. Let ρ be any irrational number between p_i and q_i. The decomposition of $\mathcal{R}_\omega$ in the closed disjoint sets determined by the inequalities:

$$x_i \leqq \rho, \qquad x_i \geqq \rho,$$

displays the sought-for separation of p and q.

Nevertheless there is equivalence in an important case:

A) For compact* spaces conditions (0)–(3) are equivalent. It re-

* Following the usage of Bourbaki, we say of a general space X that it is *compact* if from every collection of open subsets whose sum is X one can extract a finite subcollection whose sum is X. This is Alexandroff and Urysohn's "bicom-

mains to prove that (0) implies (1) and (1) implies (2). We first prove two Propositions.

B) Let X be a compact space, C a closed subset of X, and p a point of X. Then if p and each point of C can be separated, p and C can be separated.

PROOF. For each point q of C there exist two disjoint sets U_q and V_q, with $p \varepsilon U_q$ and $q \varepsilon V_q$, which are both open and closed. Since C is a closed subspace of a compact space there exist a finite number $q_1, \cdots, q_k$ of the q's such that $V_{q_1} + \cdots + V_{q_k} \supset C$. Let

$$U = \prod_{i=1}^{k} U_{q_i} \quad \text{and} \quad V = \sum_{i=1}^{k} V_{q_i}$$

Then $p \varepsilon U$ and $C \subset V$, and U and V are disjoint and both open and closed. Hence p and C are separated, proving the Proposition.

C) Let X be a compact space, p a point of X, and $M(p)$ the set of all points which cannot be separated from p. Then $M(p)$ is connected.

PROOF. We show first that $M(p)$ is closed, or what is the same, $X - M(p)$ is open. An arbitrary point x is in $X - M(p)$ if and only if there is a decomposition

$$X = U + V$$

$$UV = 0$$

$$x \varepsilon U, \qquad p \varepsilon V,$$

U and V open. It follows easily that $U \subset X - M(p)$, i.e. each point in $X - M(p)$ has a neighborhood U contained in $X - M(p)$, proving that $X - M(p)$ is open.

$M(p)$ obviously contains p. Suppose $M(p)$ were disconnected. Then

$$M(p) = C + K, \quad C \neq 0, \quad K \neq 0, \quad CK = 0,$$

C and K closed in $M(p)$, and, by notation, $p \varepsilon C$. Since $M(p)$ is closed in X, C and K are closed in X. Hence there exists, by the normality of X, a set U which is open in X such that $C \subset U$ and $\overline{U} \cdot K = 0$. Since

$$B(U) = \text{bdry } U = \overline{U} - U$$

---

pactness." No confusion can occur here because all our spaces have a countable basis, and for such spaces compactness and bicompactness, in the terminology of Alexandroff and Urysohn, have the same force.

we have

$$B(U)\cdot M(p) = B(U)(C+K)=0.$$

This means that each point of $B(U)$ is separated from p. Because $B(U)$ is closed we may apply B), to obtain a set V such that $B(U) \subset V$, $p\cdot V = 0$, V both open and closed. Because $p \in C \subset U$, $p \in U - V$. It is easy to see that we can write $U - V$ as $U - \overline{V}$, since $V = \overline{V}$, and also as $\overline{U} - V$, since $B(U) \subset V$; the first form shows that $U - V$ is open and the second shows that $U - V$ is closed. But $(U - V)K = 0$. Hence p is separated from the points of $K \subset M(p)$. However this contradicts the definition of $M(p)$.

Proof that (0) implies (1). Assume that X is totally-disconnected. Consider for each p the set $M(p)$. This is connected, by C), and therefore consists of the single point p, by the total-disconnectedness of X. Hence any two points of X are separated.

Proof that (1) implies (2). This follows from B).

D) Among compact spaces, 0-dimensional spaces are identical with totally-disconnected spaces.

Remark 1. Proposition D) holds also for spaces which are only locally compact.

Remark 2. It is *not* true (see Examples II 16 and II 17) that if a space has property (0) or (1) it will retain that property upon the adjunction of a single point; compare this with Corollary 3 to Theorem II 2. Hence the Sum Theorem would not be true for a theory of dimension in which dimension 0 were either defined by total-disconnectedness or the separation of pairs of points.

Example II 16. Knaster and Kuratowski: Sur les ensembles connexes, *Fund. Math.* 2 (1921), pp. 206–255, give (on p. 241) the following startling example of a subset of the plane which is totally-disconnected, property (0), but which loses this property, and in fact becomes connected, upon the adjunction of a single point: Denoting by $\mathcal{C}$ the Cantor discontinuum (see Example II 3), we let P be the subset of $\mathcal{C}$ made up of points

$$p = \sum_{n=1}^{\infty} a_n/3^n$$

for which a_n is either always 0 for sufficiently large n, or else a_n is always 2 for sufficiently large n. (P is the countable set made up of the end-points of the intervals which one has to remove from [0, 1] in order to get $\mathcal{C}$.) Let Q be the other points of $\mathcal{C}$. Let a be the point in the plane whose coordinates are $(\frac{1}{2}, \frac{1}{2})$ and denote by $L(c)$ the segment joining the variable point c of $\mathcal{C}$ to a. Denote by $L^*(p)$ the set of all points on $L(p)$ having *rational* ordinates, p being in P, and by $L^*(q)$ the set of all points on $L(q)$ having irrational ordinates, q being in Q. Then, as Knaster and Kuratowski prove,

$$X = \sum_{p \varepsilon P} L^*(p) + \sum_{q \varepsilon Q} L^*(q)$$

is connected, although $X - a$ is totally disconnected.

Since X is connected, X is not 0-dimensional,† and since a 0-dimensional set remains 0-dimensional after the adjunction of a single point (Corollary 3 to Theorem II 2), neither is $X - a$ 0-dimensional.

EXAMPLE II 17. Sierpiński (see the reference in Example II 14) gives an example of a space which has property (1) but loses it upon the adjunction of a single point.

† As a matter of fact (see Example IV 3),

$$\dim (X - a) = \dim X = 1.$$

$X - a$ is an example of a space having dimension 1 even though it is totally disconnected. Moreover Mazurkiewicz (Sur les problèmes κ et λ de Urysohn, *Fund. Math.* 10 (1927), pp. 311–319) has shown that for each finite n there exists a space of dimension n which is totally disconnected (in fact, has the property that any two points can be separated; see (1) on page 20). Thus the fact that a space has positive dimension implies very little about its connectedness.

CHAPTER III

Dimension n

Roughly speaking, we may say that a space has dimension $\leqq n$ if an arbitrarily small piece of the space surrounding each point may be delimited by subsets of dimension $\leqq n - 1$. This method of definition is inductive, and an elegant starting point for the induction is given by prescribing the null set as the (-1)-dimensional space.

1. Definition of dimension n

Definition III 1. The empty set and only the empty set has *dimension* -1.

A space X has *dimension* $\leqq n$ $(n \geqq 0)$ *at a point* p if p has arbitrarily small neighborhoods whose boundaries have dimension $\leqq n - 1$.

X has *dimension* $\leqq n$, $\dim X \leqq n$, if X has dimension $\leqq n$ at each of its points.

X has *dimension* n *at a point* p if it is true that X has dimension $\leqq n$ at p and it is false that X has dimension $\leqq n - 1$ at p.

X has *dimension* n if $\dim X \leqq n$ is true and $\dim X \leqq n - 1$ is false.

X has *dimension* ∞ if $\dim X \leqq n$ is false for each n.

A) It is obvious that the property of having dimension n (or of having dimension n at a point p) is topologically invariant. Dimension is, however, not an invariant of continuous transformations. Projection of a plane into a line is an illustration of a transformation which lowers dimension, and Peano's mapping of an interval on the whole of a square is an illustration of a continuous transformation which raises dimension.

B) Equivalent to the condition that $\dim X \leqq n$ is the existence of a basis of X made up of open sets whose boundaries have dimension $\leqq n - 1$.

C) For $n = 0$ it is clear that Definitions II 1 and III 1 coincide.

D) Suppose $\dim X = n$, n finite. Then X contains an m-dimensional subset for every $m \leqq n$.

Proof. Because $\dim X > n - 1$ there is a point $p_0 \in X$ and a neighborhood U_0 of p_0 with the property that if V is any open set satisfying $p_0 \varepsilon V \subset U_0$, then

$$\dim \text{bdry } V \geqq n - 1.$$

On the other hand, because $\dim X \leqq n$, there exists an open set V_0 satisfying $p_0 \varepsilon V_0 \subset U_0$ for which

$$\dim \text{bdry } V_0 \leqq n - 1.$$

Hence the boundary of V_0 is a subset of X of the precise dimension $n - 1$. The rest of Proposition D) is now evident.

Remark. The statement of Proposition D) cannot be extended to infinite-dimensional spaces. Indeed, under the hypothesis of the continuum there even exist infinite-dimensional spaces whose only finite-dimensional subspaces are countable sets.*

Example III 1. The Euclidean line, and an interval in the Euclidean line, has dimension 1.

Example III 2. Any polygon has dimension 1.

Example III 3. Any 2-manifold has dimension $\leqq 2$. (The proof follows from Example III 2.)

Example III 4. The Euclidean n-space E_n has dimension $\leqq n$. The inductive proof of this is left to the reader. The proof that the dimension of E_n is precisely n is by no means trivial, however, and will be the main concern of Chapter IV.

Example III 5. The set $\mathfrak{R}_\omega$ of points in Hilbert space all of whose coordinates are rational is one-dimensional. We have already seen (Example II 11) that $\dim \mathfrak{R}_\omega \geqq 1$. We shall now show that $\dim \mathfrak{R}_\omega \leqq 1$, by proving that each point of $\mathfrak{R}_\omega$ can be surrounded by arbitrarily small spherical neighborhoods in $\mathfrak{R}_\omega$ whose boundaries are 0-dimensional. Let us denote by S the sphere in Hilbert space of radius $d < 1$ around the origin, i.e., the set of points at distance d from the origin. It is quite clear that it suffices to prove that $\dim S \cdot \mathfrak{R}_\omega = 0$. To do this let us associate with each point x of S:

$$x = (x_1, x_2, \cdots, x_i, \cdots),$$

the point

$$x' = \left(x_1, \frac{x_2}{2}, \cdots, \frac{x_i}{i}, \cdots\right);$$

* W. Hurewicz: Une remarque sur l'hypothèse du continu, *Fund. Math.* 19 (1932), pp. 8–9.

x' is in the Hilbert-cube (see the index). As can easily be shown, S has the property that a sequence $\{p^n\}$ of points of S converges to a point p of S if and only if the ith coordinate of p^n converges to the ith coordinate of p. Since I_ω has the same property the transition from x to x' is a homeomorphism of S on a subset of I_ω. But this homeomorphism obviously sends $S \cdot \mathfrak{R}_\omega$ onto a subset of $\mathfrak{R}_\omega'$, and we know (Example II 9) that dim $\mathfrak{R}_\omega' = 0$. Hence dim $S \cdot \mathfrak{R}_\omega = 0$.

Theorem III 1. *A subspace of a space of dimension* $\leqq n$ *has dimension* $\leqq n$.

PROOF. (By induction.) The statement is obvious for $n = -1$. Assume it now for $n - 1$. Let X be a space of dimension $\leqq n$, X' a subspace of X, and p any point of X'. Let U' be a neighborhood in X' of p. Then there exists a neighborhood U in X of p such that $U' = UX'$. Because dim $X \leqq n$ there exists a set V, open in X, and satisfying

$$p \in V \subset U.$$

$$\dim \text{bdry } V \leqq n - 1.$$

Let $V' = VX'$. Then V' is open in X', $p \in V' \subset U'$. Let B be the boundary of V in X, and B' the boundary* of V' in X'. Then, as one can easily see, B' is contained in BX'. By the hypothesis of the induction, dim $B' \leqq n - 1$, q.e.d.

We now prove that Definition III 1 is equivalent to

Definition III 1'. X has dimension $\leqq n$ if every point p can be separated by a closed set of dimension $\leqq n - 1$ from any closed set C not containing p.

PROOF. Suppose dim $X \leqq n$ in the sense of Definition III 1. Now $X - C$ is a neighborhood of p and therefore, by the regularity† of X, there is another neighborhood V of p satisfying

$$\bar{V} \subset X - C.$$

* In formulas: $B = \bar{V} - V$, $B' = (\bar{V}' - V')X'$.

† A space is *regular* (AH, p. 68) if every neighborhood U of a point p contains a neighborhood V of p satisfying

$$\bar{V} \subset U.$$

Obviously every metric space is regular, and every normal space is regular (AH, p. 68).

Then there exists a neighborhood W of p for which $W \subset V$, with $B = \text{bdry } W$ of dimension $\leqq n - 1$. It is easily shown that B separates p and C. This proves that $\dim X \leqq n$ in the sense of Definition III 1′.

Conversely, if $\dim X \leqq n$ in the sense of Definition III 1′, then $\dim X \leqq n$ in the sense of Definition III 1. For let U be a neighborhood of p. Then $X - U$ is a closed set not containing p and therefore $X - U$ can be separated from p by a closed set B of dimension $\leqq n - 1$. This means that

$$X - B = U' + V', \qquad p \in U', \qquad X - U \subset V', \qquad U'V' = 0,$$

with U', V' open in $X - B$ and therefore in X. Now U' is a neighborhood of p contained in U and since the boundary of U' is contained in B, we may conclude from Theorem III 1 that the boundary of U' has dimension $\leqq n - 1$.

2. Dimension of subspaces. Dimension of sums

In considering the dimension of a subspace X' of a larger space X it is sometimes convenient to determine the dimension of X' by means of neighborhoods referring to the entire space X:

A) A subspace X' of a space X has dimension $\leqq n$ if and only if every point of X' has arbitrarily small neighborhoods, in X, whose boundaries have intersections with X' of dimension $\leqq n - 1$.

Proof. Suppose X' satisfies the conditions of A). Let p be an arbitrary point of X' and U' a neighborhood in X' of p. Then there exists a neighborhood U in X of p such that $U' = UX'$. Hence there exists a set V which is open in X such that

$$p \in V \subset U,$$
$$\dim (X' \cdot \text{bdry } V) \leqq n - 1.$$

Let $V' = VX'$. Then V' is open in X', $p \in V' \subset U'$. Again denoting by B and B' the boundaries of V in X and of V' in X' we have $B' \subset BX'$. Hence $\dim B' \leqq n - 1$, so that $\dim X' \leqq n$.

Conversely, suppose $\dim X' \leqq n$. Let p be an arbitrary point of X' and U a neighborhood in X of p. Then

$$U' = UX'$$

is a neighborhood in X' of p. Hence there is a neighborhood V' in X' of p for which

$$p \varepsilon V' \subset U', \quad \text{and}$$
$$\dim B' \leqq n - 1,$$

B' being the boundary in X' of V'. Neither of the disjoint sets V' and $X' - \overline{V}'$ contains a cluster-point of the other, so that by the complete normality* of X there exists an open set W satisfying

$$V' \subset W \quad \text{and} \quad \overline{W}(X' - \overline{V}') = 0.$$

Replacing W if necessary by $W \cdot U$ we may assume that

$$W \subset U.$$

The set

$$\overline{W} - W = \text{bdry } W$$

contains no point of $X' - \overline{V}'$ and no point of V'. It follows that the intersection of X' with the boundary of W is contained in B' and hence (Theorem III 1) has dimension $\leqq n - 1$, so that the condition of A) is fulfilled.

We now use A) to prove an extremely important proposition regarding the dimension of the sum of two sets. It has already been observed that the dimension of a sum $A + B$ is not determined by the dimensions of A and B. However:

B) For any two subspaces A, B of a space X

$$\dim (A + B) \leqq 1 + \dim A + \dim B.$$

PROOF. (By double induction on the dimensions of A and B.) The proposition is evident for

$$\dim A = \dim B = -1.$$

Now let $\dim A = m$, $\dim B = n$ and assume the proposition for the cases

(1) $\dim A \leqq m, \quad \dim B \leqq n - 1,$ and

(2) $\dim A \leqq m - 1, \quad \dim B \leqq n.$

Let p be a point of $A + B$. As a matter of notation we may take p in A. Let U be a neighborhood of p in X. By A) there exists an open set V,

$$p \varepsilon V \subset U, \quad \text{and}$$
$$\dim (W \cdot A) \leqq m - 1,$$

* See index.

where W is the boundary of V. But $W \cdot B$, as a subset of B, satisfies

$$\dim (W \cdot B) \leqq n.$$

By the hypotheses (1) and (2) of the induction

$$\dim W(A + B) \leqq m + n.$$

This proves by A) that

$$\dim (A + B) \leqq m + n + 1,$$

which completes the induction.

C) The sum of $n + 1$ subspaces each of dimension $\leqq 0$ has dimension $\leqq n$.

EXAMPLE III 6. Suppose $0 \leqq m \leqq n$. Denote by $\mathcal{M}_n^m$ the set of points in E_n at *most* m of whose coordinates are rational and by $\mathcal{L}_n^m$ the set of points in E_n at *least* m of whose coordinates are rational. Then*

$$\dim \mathcal{M}_n^m \leqq m, \quad \text{and}$$
$$\dim \mathcal{L}_n^m \leqq n - m.$$

For evidently

$$\mathcal{M}_n^m = \mathcal{R}_n^0 + \mathcal{R}_n^1 + \cdots + \mathcal{R}_n^m, \quad \text{and}$$
$$\mathcal{L}_n^m = \mathcal{R}_n^m + \mathcal{R}_n^{m+1} + \cdots + \mathcal{R}_n^n.$$

The assertion then follows from C) and the fact that each summand is 0-dimensional (Example II 12).

EXAMPLE III 7. Suppose $0 \leqq m$. Denote by $\mathcal{M}_\omega^m$ the set of points in the Hilbert cube I_ω at most m of whose coordinates are rational. Then†

$$\dim \mathcal{M}_\omega^m \leqq m.$$

* As a matter of fact

$$\dim \mathcal{M}_n^m = m$$
$$\dim \mathcal{L}_n^m = n - m.$$

See Example IV 1. Moreover we shall prove later (Theorem V 5) that every n-dimensional space can be topologically imbedded in $\mathcal{M}_{2n+1}^n$.

† As a matter of fact

$$\dim \mathcal{M}_\omega^m = m.$$

See Example IV 2.

For

$$\mathfrak{M}_{\omega}^{m} = \mathcal{R}_{\omega}^{0} + \mathcal{R}_{\omega}^{1} + \cdots + \mathcal{R}_{\omega}^{m}.$$

The assertion then follows from C) and the fact that each summand is 0-dimensional (Example II 13).

3. The sum and decomposition theorems for n-dimensional sets

We now prove the most important theorems of the abstract part of dimension theory.

Theorem III 2. The Sum Theorem for Dimension n. *A space which is the countable sum of closed subsets of dimension $\leqq n$ has dimension $\leqq n$.*

PROOF. (By induction.) Denote the sum theorem for dimension n by Σ_n. Obviously Σ_n is equivalent to the statement that any space which is the countable sum of F_σ sets of dimension $\leqq n$ has dimension $\leqq n$.

Σ_{-1} is trivial. We shall now deduce Σ_n from Σ_{n-1}, making use of Σ_0, which has already been proved independently (Theorem II 2).

First we prove that Σ_{n-1} implies the following proposition:

Δ_n. Any space of dimension $\leqq n$ is the sum of a subspace of dimension $\leqq n - 1$ and a subspace of dimension $\leqq 0$.

PROOF OF Δ_n. Let X be a space of dimension $\leqq n$. Then by Proposition 1B) there exists a basis for the open sets of X made up of sets whose boundaries have dimension $\leqq n - 1$. Since X is separable there exists* a countable basis $\{U_i\}$, $i = 1, 2, \cdots$, made up of sets whose boundaries $\{B_i\}$ have dimension $\leqq n - 1$. From Σ_{n-1} it follows that

$$B = \sum_{i=1}^{\infty} B_i$$

has dimension $\leqq n - 1$.

We assert that

$$\dim (X - B) \leqq 0. \tag{1}$$

For obviously the boundaries of the sets U_i do not meet $X - B$, and hence the condition of Proposition 2A) (with $n = 0$ and X' replaced by $X - B$) is satisfied. Δ_n then follows from the equation $X = B + (X - B)$.

* AH, p. 78.

Now we shall combine Σ_{n-1} and Δ_n to prove Σ_n. Suppose

$$X = C_1 + \cdots + C_i + \cdots,$$
$$\dim C_i \leqq n,$$

each C_i closed. We want to show that $\dim X \leqq n$. Let

$$K_1 = C_1,$$
$$K_i = C_i - \sum_{j=1}^{i-1} C_j = C_i\left(X - \sum_{j=1}^{i-1} C_j\right), \quad i = 2, 3, \cdots.$$

Then

$$X = \sum_{i=1}^{\infty} K_i, \tag{2}$$

$$K_iK_j = 0 \quad \text{if} \quad i \neq j, \tag{3}$$

$$K_i \text{ is an } F_\sigma \text{ in } X, \tag{4}$$

$$\dim K_i \leqq n. \tag{5}$$

(2) and (3) are obvious. To prove (4) note that

$$\sum_{j=1}^{i-1} C_j$$

is closed; hence

$$X - \sum_{j=1}^{i-1} C_j$$

is open, and, therefore, as an open set in a metric space is* an F_σ; K_i, as the intersection of this F_σ with the closed set C_i, is thus also an F_σ. (5) holds because K_i is a subset of C_i.

(5) enables us to apply Δ_n to each K_i: we have

$$K_i = M_i + N_i,$$
$$\dim M_i \leqq n - 1, \qquad \dim N_i \leqq 0.$$

Denote $\sum M_i$ by M and $\sum N_i$ by N. From (2),

$$X = M + N.$$

Each M_i is an F_σ in M. For

$$M_i = M_iK_i = (M_1 + \cdots + M_i + \cdots)K_i = MK_i,$$

since $M_i \subset K_i$ and $K_iK_j = 0$ for $i \neq j$ by (3). Hence M_i, as the in-

* Kuratowski, *Topologie* I, p. 87.

tersection of M with K_i, which is an F_σ by (4), is itself an F_σ in M, Therefore we may apply Σ_{n-1} to conclude that dim $M \leqq n - 1$. By a similar argument each N_i is an F_σ in N and therefore dim $N \leqq 0$ by* Σ_0.

Thus we have $X = M + N$ with dim $M \leqq n - 1$ and dim $N \leqq 0$. From 2 B) we conclude that dim $X \leqq n$, q.e.d.

Corollary 1. *The sum of two subspaces each of which has dimension $\leqq n$ and one of which is closed has dimension $\leqq n$.*

Proof. As in Corollary 2 to Theorem II 2.

Corollary 2. *The dimension of a non-empty space cannot be increased by the adjunction of a single point.*

Proof. Corollary 2 is an obvious consequence of Corollary 1.

Corollary 3. *If a space X' of dimension $\leqq n$ is contained in an arbitrary space X then every point of the containing space has arbitrarily small neighborhoods (in X) whose boundaries have intersections with X' of dimension $\leqq n - 1$.* (Compare this with 2 A) and observe that 2 A) imposes a condition on the neighborhoods of points of X' only.)

Proof. For each point p of X, $X' + p$ has dimension $\leqq n$ by Corollary 2; the statement thus follows from 2 A).

Corollary 4. *If a space has dimension $\leqq n$ it is the sum of a subspace of dimension $\leqq n - 1$ and a subspace of dimension $\leqq 0$.*

Proof. This is Δ_n, which we have shown in the proof of Theorem III 2 to be a consequence of Σ_{n-1}.

Theorem III 3. The Decomposition Theorem for Dimension n. *A space has dimension $\leqq n$, n finite, if and only if it is the sum of $n + 1$ subspaces of dimension $\leqq 0$.*

Proof. Theorem III 3 follows from repeated application of Corollary 4 above, and 2 C).

Corollary. *If* dim $X = n$ *and p, q are two integers $\geqq -1$ such*

* It is the explicit use made here of Σ_0 which necessitated the separate proof, in Theorem II 2, of Σ_0.

that $p + q + 1 = n$, then X is the sum of two subspaces P and Q of dimensions p and q respectively.

Proof. Directly from Theorem III 3.

Example III 8. We have already exhibited (Example II 12) a decomposition of E_n into the $n + 1$ spaces $\mathcal{R}_n^0, \cdots, \mathcal{R}_n^n$ of dimension 0.

4. Dimension of a topological product

Theorem III 4. The Product Theorem. *Let us denote by $A \times B$ the topological product* of two spaces A and B, at least one of which is nonempty. Then*

$$\dim (A \times B) \leqq \dim A + \dim B.$$

Proof. (By induction.) The proposition is evident if either

$$\dim A = -1 \quad \text{or} \quad \dim B = -1.$$

Now let $\dim A = m$, $\dim B = n$ and assume the proposition for the cases

$$\dim A \leqq m, \qquad \dim B \leqq n - 1, \quad \text{and} \tag{1}$$

$$\dim A \leqq m - 1, \qquad \dim B \leqq n. \tag{2}$$

Each point $p = (a, b)$ in $A \times B$ has arbitrarily small neighborhoods of the form $U \times V$, U being a neighborhood of a in A and V a neighborhood of b in B, and we may assume that

$$\dim \text{bdry } U \leqq m - 1, \qquad \dim \text{bdry } V \leqq n - 1.$$

Now

$$\text{bdry } (U \times V) = \overline{U} \times \text{bdry } V + \overline{V} \times \text{bdry } U.$$

Each summand is closed, and by hypotheses (1) and (2) of the induction has dimension $\leqq m + n - 1$. Hence, by the Sum Theorem,

$$\dim \text{bdry } (U \times V) \leqq m + n - 1,$$

which proves that

$$\dim (A \times B) \leqq m + n.$$

Corollary. *If B is 0-dimensional then*

$$\dim (A \times B) = \dim A + \dim B. \tag{3}$$

* See index.

PROOF. Since B is not empty

$$A \times B \supset A.$$

Therefore

$$\dim (A \times B) \geqq \dim A = \dim A + \dim B;$$

combining this with Theorem III 4 proves the Corollary.

REMARK. One might expect the logarithmic law (3) to be true in general. Unfortunately this is not so, for it is clear that $\mathcal{R}_\omega$, the set of points in Hilbert-space all of whose coordinates are rational, is homeomorphic to $\mathcal{R}_\omega \times \mathcal{R}_\omega$, while Example III 5 shows that $\dim \mathcal{R}_\omega = 1$. (3) does not hold even if both A and B are compact. This is shown by Pontrjagin's example (*Comptes Rendus*, 190 (1930), pp. 1105–1107) of two compact 2-dimensional spaces whose product is 3-dimensional. It can be shown that (3) holds if B is one-dimensional, provided A is compact. It is an open problem to characterize the spaces B for which (3) holds for arbitrary A.

5. Separation of sets in n-dimensional spaces

A) If a space X has dimension $\leqq n$ then any two disjoint closed subsets in X can be separated by a closed set of dimension $\leqq n - 1$.

Proposition A) is included (for $X = A$) in the more general proposition:

B) If C_1 and C_2 are disjoint closed sets in a space X (of arbitrary dimension) and A is a subset of X of dimension $\leqq n$, then there exists a closed set B separating C_1 and C_2 with $\dim AB \leqq n - 1$.

PROOF. If $n = 0$ either $\dim A = -1$ and B) is obvious, or else $\dim A = 0$, in which case B) has already been demonstrated in II 2F).

Suppose then that $n > 0$. Applying Corollary 4 of Theorem III 2 to A we have $A = D + E$, with $\dim D \leqq n - 1$, $\dim E \leqq 0$. Let us use Proposition B) for $n = 0$ to obtain a separation of C_1 and C_2 by a set B not meeting E. Hence

$$AB \subset D.$$

But $\dim D \leqq n - 1$; therefore $\dim AB \leqq n - 1$, q.e.d.

REMARK. The converse of A) is obvious, so that A) establishes an equivalence between the property in the small of separation of a point from a closed set and the property in the large of the separation of two closed sets. It is this conjunction of properties in the small and the large which is largely responsible for the power of the dimension-concept (see the Appendix).

The following proposition will play an important role in Chapter IV:

C) Let X be a space of dimension $\leqq n - 1$ and let C_i, C_i', $i = 1, \cdots, n$, be n pairs of closed subsets of X for which

$$C_iC_i' = 0.$$

Then there exist n closed sets B_i such that B_i separates C_i and C_i', and

$$B_1B_2 \cdots B_n = 0.$$

Proof. From A) we get a closed set B_1 separating C_1 and C_1' with

$$\dim B_1 \leqq n - 2.$$

Using B) we get a closed set B_2 separating C_2 and C_2' with

$$\dim B_1B_2 \leqq n - 3.$$

By repeating this application of B) we arrive at n sets B_i such that B_i separates C_i and C_i' and

$$\dim B_1B_2 \cdots B_k \leqq n - k - 1, \qquad k = 1, \cdots, n.$$

For $k = n$ we conclude that $B_1B_2 \cdots B_n = 0$.

Remark. The property of Proposition C) is *characteristic* for spaces of dimension less than n (see the Remark on page 78).

6. Compact spaces

Consider the following properties of a space X:

(1) Any two distinct points can be separated by a closed set of dimension $\leqq n - 1$.

(2) Any point can be separated from a closed set not containing it by a closed set of dimension $\leqq n - 1$, i.e. $\dim X \leqq n$ (see Definition III 1′).

(3) Any two disjoint closed sets can be separated by a closed set of dimension $\leqq n - 1$.

It is clear that (3) implies (2) and (2) implies (1). We have already proved (5 A)) that (2) implies (3) for arbitrary (separable metric) spaces, and in Section 4 of Chapter II we discussed the relations between these properties in the case $n = 0$. We now prove, in analogy with II 4 A), that

A) For compact spaces properties (1), (2), and (3) are equivalent.

PROOF. It remains to prove that (1) implies (2). We do this by demonstrating

B) Let X be a compact space, C a closed subset of X and p a point of X. Then if p can be separated from each point of C by a closed set of dimension $\leqq n - 1$, p can be separated from C by a closed set of dimension $\leqq n - 1$.

PROOF. For each point q of C we have an open $U(q)$ for which

$$q \varepsilon U(q), \qquad p \notin \overline{U(q)},$$
$$\dim \operatorname{bdry} U(q) \leqq n - 1.$$

C, as a closed subset of a compact space, is compact, and therefore there exists a finite number $q_1, \cdots, q_k$ of the points q such that

$$C \subset U = U(q_1) + \cdots + U(q_k).$$

Suppose B is the boundary of U, $B = \overline{U} - U$. Then

$$B \subset \operatorname{bdry} U(q_1) + \cdots + \operatorname{bdry} U(q_k),$$

and thus

$$\dim B \leqq n - 1,$$

by the Sum Theorem for dimension $n - 1$. Moreover $p \notin \overline{U}$. Hence p is separated from C by the closed set B of dimension $\leqq n - 1$, q.e.d.

C) A compact space has dimension $\leqq n$ if and only if any two distinct points in it can be separated by a closed set of dimension $\leqq n - 1$.

REMARK. C) holds also for spaces which are only locally compact.

CHAPTER IV

Dimension of Euclidean Spaces

In this chapter our concept of dimension will be justified by the proof that the dimension of Euclidean n-space is precisely n.

The reader will observe that so far not even the existence of spaces of dimension > 1 has been shown.

1. Some topological properties of E_n

We already know (Example III 4) that dim $E_n \leqq n$, and it remains to prove that dimension $E_n \geqq n$. Before we do this we need some general theorems on spheres and Euclidean spaces.

It is intuitively clear, at least for $n = 1$ and $n = 2$, that an n-sphere cannot be shrunk to a point in itself;* by this we mean that it is not possible to displace the points of an n-sphere S_n in a finite time-interval along paths which lie in S_n and have a common end point, the position of a moving point depending continuously on the time and the original position of the point. Formally:

A) Let S_n be an n-sphere, e.g. the set of all points in E_{n+1} at distance 1 from the origin. Then there exists no function $f_t(x)$ of the two variables $x \in S_n$ and $0 \leqq t \leqq 1$, with values in S_n, continuous in the pair (x, t), and with the boundary conditions

(1) $$f_0(x) = x \quad \text{and} \quad f_1(x) = \text{constant.}\dagger$$

For the reader who is familiar with the concept of the degree of a mapping (of an n-sphere in itself) and the result that the degree of a mapping is invariant under continuous deformation we may proceed very quickly: the identity mapping f_0 has the degree 1, the constant mapping f_1 the degree 0; hence it is impossible that f_0 and f_1 belong to the same continuous family f_t.

The following is a complete elementary proof.

* It is of course trivial to shrink a 2-sphere to a point in Euclidean 3-space. On the other hand, a rubber membrane covering a solid ball cannot be shrunk to a point without tearing it, and this is exactly what is intuitively meant by the non-shrinkability of the 2-sphere.

† f_1 is a "constant" mapping; in general, a mapping f of a space X in a space Y is called a *constant* mapping if f sends all of X into a single point of Y.

PROOF. We use the simple geometrical fact that if $p_0, \cdots, p_n$ are $n + 1$ points on the unit sphere S_n whose distances in pairs are less than 1, then $p_0, \cdots, p_n$ determine a unique (perhaps degenerate) spherical n-simplex.*

(i) Let T be a triangulation of S_n in spherical n-simplexes. Suppose we assign to each vertex a_i of T a point $\varphi(a_i) \varepsilon S_n$ in such a way that given any n-simplex Δ of T, the points assigned to its vertices form a set of diameter less than 1, and therefore determine a new n-simplex Δ^φ (which may be degenerate). φ thus associates with the triangulation T of S_n a collection T^φ of spherical n-simplexes of S_n; the simplexes T^φ will in general overlap. For any point $p \varepsilon S_n$ not on the boundary of a Δ^φ of T^φ we denote by

$$n(p, \varphi, T)$$

the number of n-simplexes Δ^φ containing p. We maintain that if q is another point of S_n not on the boundary of a Δ^φ of T^φ then

$$n(p, \varphi, T) \equiv n(q, \varphi, T) \bmod 2. \tag{2}$$

In the proof of (2) we may assume that

$$\text{the points } \varphi(a_i) \text{ are in general position,}\dagger \tag{3}$$

i.e. if $m \leqq n$ no $m + 1$ of the $\varphi(a_i)$ lie on a common m-plane through the origin, since for fixed p and q we can replace the points $\varphi(a_i)$ by points $\varphi'(a_i)$ satisfying (3) which are so close to $\varphi(a_i)$ that

$$p \,\varepsilon\, \Delta_i^{\varphi} \quad \text{if and only if} \quad p \,\varepsilon\, \Delta_i^{\varphi'} \quad \text{and}$$
$$q \,\varepsilon\, \Delta_i^{\varphi} \quad \text{if and only if} \quad q \,\varepsilon\, \Delta_i^{\varphi'}, \quad \text{i.e.}$$
$$n(p, \varphi, T) = n(p, \varphi', T) \quad \text{and} \quad n(q, \varphi, T) = n(q, \varphi', T).$$

Let us draw an arc on S_n from p to q which cuts each $(n - 1)$-face of a simplex of T^φ in a finite number of points and contains no point common to more than one $(n - 1)$-face. This is possible because of (3). As the point x moves along this arc from p to q the integer $n(x, \varphi, T)$ changes only if x crosses an $(n - 1)$-face of an n-simplex Δ^φ of T^φ. Let such a face be determined by the points $\varphi(a_1), \cdots, \varphi(a_n)$. The vertices $a_1, \cdots, a_n$ determine an $(n - 1)$-simplex of the original

* The *spherical n-simplex* determined by $p_0, \cdots, p_n$ is the projection on S_n from the origin of the least convex set in E_{n+1} containing $p_0, \cdots, p_n$. This n-simplex is *degenerate* if there is an n-dimensional hyperplane containing $p_0, \cdots, p_n$ and the origin. We are not concerned with questions of orientation.

† This implies that none of the simplexes Δ^φ is degenerate.

triangulation T and this $(n - 1)$-simplex is the common boundary of exactly two n-simplexes Δ and Γ of T. Consider Δ^{φ} and Γ^{φ}. Either Δ^{φ} overlaps Γ^{φ} or Δ^{φ} does not overlap Γ^{φ}. In the first case $n(x, \varphi, T)$ changes by 2 when x crosses the $(n - 1)$-face and in the second case $n(x, \varphi, T)$ remains fixed; in either case, the *parity* of $n(x, \varphi, T)$ remains fixed. This proves (2).

We now denote by $n(\varphi, T)$ the common value mod 2 of $n(p, \varphi, T)$.

(ii) Now suppose f is a mapping* of S_n in itself. Then there exists a triangulation T of S_n in simplexes so small that the image under f of any simplex of T has diameter less than 1. We then substitute $f(a_i)$ for $\varphi(a_i)$ in (i) and thereby define $n(p, f, T)$ and $n(f, T)$.

It is clear that if f is the identity mapping, $n(f, T) = 1$, and if f is a constant mapping, $n(f, T) = 0$.

(iii) We now return to the proof of Proposition A). Suppose that a continuous function $f_t(x)$ of the type described in A) did exist. From the compactness of the ranges of x and t it follows that $f_t(x)$ is uniformly continuous in (x, t). Hence there exists a triangulation T_0 of S_n with this property: For each t, the image under f_t of a simplex of T_0 has diameter less than 1. It is clear that if p is any point for which $n(p, f_t, T_0)$ is defined there is a $\delta > 0$ such that $n(p, f_{t'}, T_0)$ is also defined and

$$n(p, f_t, T_0) = n(p, f_{t'}, T_0)$$

for every t' satisfying $|t - t'| < \delta$. Hence $n(f_t, T_0) = n(f_{t'}, T_0)$ for $|t - t'| < \delta$; this proves that $n(f_t, T_0)$ is constant. It is therefore impossible that $f_t(x)$ satisfy the boundary conditions (1), because this would imply $n(f_0, T_0) = 1$ and $n(f_1, T_0) = 0$.

REMARK. It can easily be shown that the number $n(f, T)$ used in the proof of A) depends only on f, i.e. $n(f, T) = n(f, T')$ whenever T and T' are triangulations for which the symbols have meaning. The common value $n(f)$ of $n(f, T)$ is called the *degree* mod 2 of f. Because any two triangulations T and T' have a common refinement it suffices to prove that $n(f, T) = n(f, T')$ when T' is a subdivision of T. This is almost immediate if f is a simplicial mapping with respect to T, that is if each simplex of T is mapped barycentrically† on some simplex of S_n. The case of a general mapping is reduced to this case because f can be approximated with arbitrary accuracy by a simplicial map-

* Note that throughout this book "mapping" always means a continuous transformation.

† A barycentric mapping f of a spherical simplex $s_n = (p_0, \cdots, p_n)$ on a spherical simplex $f(s_n) = (f(p_0), \cdots, f(p_n))$ is one that assigns to each $p = \sum_{i=0}^{n} a_i p_i$ of s_n the point $f(p) \in f(s_n)$ that is the projection of $\sum_{i=0}^{n} a_i f(p_i)$ on S_n from the origin.

ping f' with T sufficiently fine. The argument of (iii) shows that $n(f, T) = n(f', T)$.

Proposition A) can also be formulated as follows:

B) Let K_n be a closed spherical region of E_n, e.g. the set of points in E_n whose distance from the origin is less than or equal to 1, and let S_{n-1} be the $(n-1)$-sphere which is its boundary. Then there exists no mapping F of K_n in S_{n-1} which keeps each point of S_{n-1} fixed.

PROOF. For suppose F were such a mapping. Then the function

$$f_t(x) = F((1-t)x),\ x \varepsilon S_{n-1},$$

would satisfy (1) (we here regard x as a unit vector in E_n emanating from the origin; $(1-t)x$ is the vector whose components are the components of x multiplied by $1-t$).

A consequence of B) is one of the best known theorems of topology:

C) **The Brouwer Fixed-Point Theorem.** *A mapping of a closed spherical region K_n of E_n in itself always has a fixed point, i.e. a point coinciding with its image.*

PROOF. Suppose a continuous function $g(x)$ defined over K_n did exist with $g(x) \neq x$ for every x. Let S_{n-1} be the $(n-1)$-sphere which is the boundary of K_n. Consider for each $x \varepsilon K_n$ the ray joining x and $g(x)$, directed from $g(x)$ to x, and let $f(x)$ be the intersection of this ray with S_{n-1}. Then the correspondence

$$x \rightarrow f(x)$$

is evidently a mapping of K_n in S_{n-1} which keeps each point of S_{n-1} fixed. This, however, is impossible, because it contradicts B).

D) Let I_n be a cube in E_n, e.g. the set of points each of whose n coordinates $x_1, \cdots, x_n$ satisfies

$$|x_i| \leqq 1.$$

Let C_i be the face of I_n determined by the equation

$$x_i = 1$$

and C_i' the face opposite. Let B_i be a closed set separating C_i and C_i'. Then

$$B_1 \cdots B_n \neq 0.$$

PROOF. B_i separates C_i and C_i' in I_n, i.e.

$$I_n - B_i = U_i + U_i',$$
$$C_i \subset U_i, \qquad C_i' \subset U_1',$$
$$U_i \cdot U_i' = 0,$$

with U_i and U_i' open in $I_n - B_i$, and therefore in I_n. For each point x of I_n let $v(x)$ be the vector whose i^{th} component has the value

$$\pm\, d(x, B_i)$$

($d(x, B_i)$ denotes as usual, the distance from x to B_i), the sign being taken as $+$ if $x \varepsilon U_i'$ and as $-$ if $x \varepsilon U_i$. We place the vector $v(x)$ with its initial point at x, and assign to x as its image $f(x)$ the end-point of this vector. The rule for the sign assures that in any case $f(x) \varepsilon I_n$. The correspondence $f(x)$ is easily seen to be continuous: $f(x)$ is a mapping of I_n in itself. From the Brouwer Fixed-Point Theorem (C)), which we apply to I_n as a homeomorph of K_n, we deduce the existence of an x^0 such that

$$f(x^0) = x^0.$$

This means that

$$d(x^0, B_i) = 0$$

for each i; in other words, $x^0 \varepsilon B_i$. Hence

$$B_1 \cdot \cdot \cdot B_n \neq 0, \quad \text{q.e.d.}$$

Remark. Propositions A), B), C), and D) are interrelated closely and given any one the others follow by quite simple arguments.

2. Dimension of E_n

We now prove the most important results of this chapter.

A) dim $I_n \geqq n$.

Proof. Suppose, to the contrary that dim $I_n \leqq n-1$. Then by III 5 C) there would exist n closed subsets B_i of I_n, each separating a different pair of opposite faces with $B_1B_2 \cdot \cdot \cdot B_n = 0$. But this contradicts 1 D).

B) dim $E_n \geqq n$.

Proof. For I_n is a subset of E_n.

Theorem IV 1. *Euclidean n-space has dimension n.**

* Proved first by Brouwer: Über den natürlichen Dimensionsbegriff, *Jour. f. Math.* 142 (1913), pp. 146–152. Brouwer as well as Menger and Urysohn base their proofs on Lebesgue's covering theorem (see below).

PROOF. Combine B) with Example III 4.

COROLLARY. *The Euclidean n-cube has dimension n.*

PROOF. Combine A) with the fact that I_n is a subset of E_n.

EXAMPLE IV 1. dim $\mathcal{M}_n^m = m$ and dim $\mathcal{L}_n^m = n - m$. For evidently (see Example III 6)

$$E_n = \mathcal{M}_n^m + \mathcal{L}_n^{m+1}$$

and dim $\mathcal{M}_n^m \leqq m$ and dim $\mathcal{L}_n^{m+1} \leqq n - m - 1$. Consequently unless dim $\mathcal{M}_n^m = m$ and dim $\mathcal{L}_n^{m+1} = n - m - 1$ we would have (III 2 B)) dim $E_n < n$.

EXAMPLE IV 2. dim $\mathcal{M}_\omega^m = m$. For consider the set M of points x in the Hilbert cube,

$$x = (x_1, x_2, \cdots, x_i, \cdots),$$

whose first m coordinates are unrestricted, while for all indices $k > m$, x_k has the irrational value $1/(k\pi)$. M is homeomorphic to E_m. Hence (Theorem IV 1), dim $M = m$. But M is a subset of $\mathcal{M}_\omega^m$, so that dim $\mathcal{M}_\omega^m \geqq m$. Combining this with Example III 7, we conclude that dim $\mathcal{M}_\omega^m = m$.

3. Lebesgue's covering theorem

Theorem IV 2. Lebesgue's Covering Theorem. *Suppose an n-dimensional cube is the sum of a finite number of closed sets, none of which contains points of two opposite faces. Then at least n + 1 of these closed sets have a common point.*

Although this theorem is not used in this book we prove it in view of its great historical importance (see section 3 of Chapter I). The proof is based essentially on Proposition 1 D).

First we prove two auxiliary Propositions:

A) Let A be a closed set in a space X, C and C' a pair of disjoint closed subsets of X and K a closed subset of A which separates AC and AC' in A. Then there exists a closed set B separating C and C' in X such that $AB \subset K$.

PROOF. The statement that K separates AC and AC' in A means that

$$A - K = D + D'$$
$$AC \subset D, \qquad AC' \subset D',$$
$$DD' = 0,$$

D, D' closed in $A - K$. Hence neither of the disjoint sets $C + D$ and $C' + D'$ contains a cluster-point of the other. By complete normality* there is an open set W in X such that $C + D \subset W$ and $\overline{W}(C' + D') = 0$. It follows that $B = \text{bdry } W$ separates C and C' in X and $B(D + D') = 0$; hence $AB \subset K$.

B) Let C_i and C_i', $i = 1, \cdots, n$, be the pairs of opposite faces of I_n. Suppose

(1) $$K_1 \supset K_2 \supset \cdots \supset K_n$$

is a decreasing sequence of closed sets of I_n such that K_1 separates C_1 and C_1' in I_n, K_2 separates K_1C_2 and K_1C_2' in $K_1, \cdots, K_n$ separates $K_{n-1}C_n$ and $K_{n-1}C_n'$ in K_{n-1}. Then K_n is not empty.

Proof. Let $B_1 = K_1$; then use A) to extend each K_i $(i = 2, \cdots, n)$ to a set separating C_i and C_i' in I_n, taking A as K_{i-1} and C and C' as C_i and C_i'. This gives us a collection $B_1, \cdots, B_n$ of closed sets for which

(2) $$B_i \text{ separates } C_i \text{ and } C_i' \text{ in } I_n,$$

(3) $$B_1 = K_1, \qquad B_iK_{i-1} \subset K_i, \qquad i = 2, \cdots, n$$

It follows from (1) and (3) that

(4) $$B_1 \cdots B_n \subset K_n;$$

it follows from (2) and Proposition 1 D) that

(5) $$B_1 \cdots B_n \neq 0.$$

Hence K_n is not empty.

Proof of Theorem IV 2. Denote by L_1 the sum of the sets in the given decomposition of I_n which meet C_1, L_2 the sum of the remaining sets which meet C_2, L_3 the sets not in $L_1 + L_2$ which meet $C_3, \cdots$, and L_{n+1} the sum of the sets which meet no C_i. Let

$$K_1 = L_1(L_2 + \cdots + L_{n+1}),$$
$$K_2 = L_1L_2(L_3 + \cdots + L_{n+1}),$$
$$K_{n-1} = L_1L_2 \cdots L_{n-1}(L_n + L_{n+1}),$$
$$K_n = L_1L_2 \cdots L_nL_{n+1}.$$

* See index.

Then it follows from the hypothesis of the theorem, as the reader may easily prove, that the sets K_i fulfill the requirements of B). Hence $K_n \neq 0$, i.e.

$$L_1 L_2 \cdots L_{n+1} \neq 0.$$

The theorem itself then follows from the fact that each of the original closed sets of the given decomposition is contained in only one L_i.

Remark. We shall see in Chapter V that Lebesgue's Theorem expresses a property common to all spaces of dimension $\geqq n$.

4. Subsets of E_n

Suppose N is a subset of E_n which contains a non-empty open set. It is clear that dim $N = n$. For there is a point p of N and a positive real number ρ such that the spherical neighborhood $S(p, \rho)$ of p of radius ρ is entirely contained in N, and $S(p, \rho)$ is clearly homeomorphic to E_n.

We shall now prove the converse:

Theorem IV 3. *A necessary and sufficient condition that a subset N of E_n be n-dimensional is that N contain a non-empty subset which is open in E_n.*

Proof. We have to show that if dim $N = n$ then N contains a non-empty open set. This is equivalent, setting M equal to the complement of N, to the statement that if M is a dense subset of E_n, then the complement of M has dimension $\leqq n - 1$. There is no loss of generality in taking M as a countable set. For M, as a subset of a separable space, contains a countable dense subset A, and dim $(E_n - A) \leqq n - 1$ implies, of course, dim $(E_n - M) \leqq n - 1$.

The statement is certainly correct (Example III 6) if we take M to be the countable set $\mathcal{R}_n$ of the points of E_n all of whose coordinates are rational. In order to conclude from this that our statement is true for an arbitrary dense countable set M it suffices to prove the homogeneity property of Euclidean spaces expressed in the following Proposition.

A) *For any two countable dense subsets A and B of E_n there exists a homeomorphism of E_n on itself which sends A onto B.*

Before we prove Proposition A) we make some preparatory remarks.

Let (x^1, x^2) and (y^1, y^2) be two ordered pairs of points of E_n. If the coordinate axes are in general position, i.e. if no parallel to a coordinate hyperplane contains more than one point x^1 or x^2 or more than one point y^1 or y^2, then we say that (x^1, x^2) and (y^1, y^2) are similarly placed if the vectors $x^1 - x^2$ and $y^1 - y^2$ are contained in the same "quadrant" of E_n, i.e. if for each $i = 1, \cdots, n$ the real numbers $x_i^1 - x_i^2$ and $y_i^1 - y_i^2$ have the same sign, x_i and y_i being the coordinates of x and y.

Let $X = x^1, x^2, \cdots, x^i, \cdots$ and $Y = y^1, y^2, \cdots, y^i, \cdots$, be two countable sequences of points or else two finite sets of the same cardinal number. It is always possible to choose the coordinate system so that the axes are in general position (no parallel to a coordinate hyperplane contains more than one point x or one point y) since this condition requires the axes to avoid at most a countable number of directions. Assuming this, we say that X and Y are similarly placed if the two ordered pairs (x^{μ_1}, x^{μ_2}) and (y^{μ_1}, y^{μ_2}) are similarly placed for each index pair μ_1, μ_2.

B) Let A and B be two countable dense sets in E_n and let the coordinate axes be in general position with respect to A and B. Then A and B may be rearranged into similarly-placed sequences.

Proof of B). Let A and B be ordered arbitrarily: $A = a^1, \cdots, a^i, \cdots$ and $B = b^1, \cdots, b^i \cdots$. We shall define sequences $C = c^1, \cdots, c^i, \cdots$ and $D = d^1, \cdots, d^i, \cdots$ which are similarly placed and which are rearrangements of A and B. This construction will be inductive and each step of the induction will consist of selecting (1) an element of C from A, (2) an element of D from B, (3) another element of D from B, and finally (4) an element of C from A.

We start by setting $c^1 = a^1$ and $d^1 = b^1$. Then we take $d^2 = b^2$ and $c^2 = a^\sigma$, where σ is the least integer such that (c^1, a^σ) and (d^1, d^2) are similarly placed. c^2 exists because the sequence A is dense in E_n.

Suppose that $c^1, \cdots, c^{2j}$ and $d^1, \cdots, d^{2j}$ have been so chosen that they are similarly placed. Now denote by c^{2j+1} the first a not yet included among the $c^1, \cdots, c^{2j}$, and by d^{2j+1} the first b such that $c^1, \cdots, c^{2j}, c^{2j+1}$ and $d^1, \cdots, d^{2j+1}$ are similarly placed. d^{2j+1} exists because B is dense. Now denote by d^{2j+2} the first b not yet included among the $d^1, \cdots, d^{2j}, d^{2j+1}$, and c^{2j+2} the first a such that $c^1, \cdots, c^{2j}, c^{2j+1}, c^{2j+2}$ and $d^1, \cdots, d^{2j}, d^{2j+1}, d^{2j+2}$ are similarly placed. c^{2j+2} exists because A is dense.

This completes the induction. It is clear that C and D are similarly placed, and that C includes every element of A and D every element of B, i.e. C and D are rearrangements of A and B. This proves Proposition B).

We now turn to the

Proof of A). Proposition B) enables us to consider the countable dense sequences A and B of the lemma as similarly placed. The lemma will be proved by extending the one-to-one correspondence

$$f: \quad a^i \rightleftarrows b^i$$

of A and B to a homeomorphism of E_n on itself.

Let $x = (x_1, \cdots, x_n)$ be an arbitrary point of E_n, not in A. We shall define $y = f(x)$ by determining for each k, $k = 1, \cdots, n$, the kth coordinate y_k of y. The points of A fall into two disjoint classes according as their k^{th} coordinates are less than or equal to x_k or greater than x_k. Associated with this decomposition of A is a decomposition of B in two disjoint classes by virtue of the one-to-one character of the correspondence $f(A) = B$. The decomposition of B induces a decom position into two disjoint classes of the set K of all the kth coordinates of elements of B. Because A and B are similarly placed, this decomposition of K has the characteristic that each element of one class is less than each element of the other class. Since K is dense in the real number continuum, this cut defines a real number, which we take to be y_k. We leave to the reader the easy proof that the mapping f so extended is a homeomorphism of E_n onto itself.* This proves A) and hence Theorem IV 3.

COROLLARY 1. *A necessary and sufficient condition that a subset N of an n-dimensional manifold (a connected space each of whose points has a neighborhood homeomorphic to E_n) be n-dimensional is that N contain a non-empty open subset of the manifold.*

COROLLARY 2. *Let U be an open set in E_n which is neither empty nor dense,† and let B be the boundary of U. Then* $\dim B = n - 1$.

PROOF. First we note from Theorem IV 3 that $\dim B \leqq n - 1$, since B contains no non-empty open set. We shall now show that it is impossible for B to have dimension $\leqq n - 2$.

Suppose first that U is bounded. Let p be a point in U. If B did have dimension $\leqq n - 2$ we could obtain, by the simple process of shrinking, arbitrarily small neighborhoods of p, homeomorphic to U, whose boundaries, homeomorphic to B, had dimension $\leqq n - 2$, i.e. E_n would have dimension $\leqq n - 1$ at the point p. On grounds of homogeneity, this would imply that E_n had dimension $\leqq n - 1$, contradicting Theorem IV 1. Hence $\dim B = n - 1$.

* It is interesting to observe that this homeomorphism has the property that the kth coordinate of the image point depends only on the kth coordinate of the original point.

† I.e., both U and its complement contain a non-empty open set.

Now suppose U is not bounded. In this case we denote by p a point in the interior of the complement of U and let ρ be a real number so small that the spherical neighborhood $S(p, \rho)$ of radius ρ about p is contained in the complement of U. We now invert E_n with respect to the center p and radius ρ. This inversion is, of course, a homeomorphism of $E_n - p$ on itself. It carries U into a non-empty bounded open set U' entirely contained in $S(p, \rho)$, and if we denote by B' the boundary of U' we have that $B' - p$ is homeomorphic to B. From the paragraph immediately above we know that $\dim B' = n - 1$. The fact that the dimension of a non-empty set cannot be increased by adjunction of a single point (Corollary 2 to Theorem III 2) proves that $\dim B = n - 1$.

Remark. Let U be an open set in an n-dimensional manifold which is neither empty nor dense, and let B be the boundary of U. Then $\dim B = n - 1$. This follows from Corollary 1 to Theorem IV 4.

Example IV 3. Let $X - a$ be the totally disconnected set of Knaster and Kuratowski which becomes connected after adjunction of the point a (see Example II 16). Then

$$\dim (X - a) = \dim X = 1.$$

For we already know that $X - a$ and X have dimension $\geqq 1$, and neither $X - a$ nor X, which are subsets of E_2, contains an open set in E_2.

5. Disconnecting sets in E_n

Definition IV 1. A subset D of a space X is said to *disconnect* X if $X - D$ is disconnected.

A) The following three statements about a space X are equivalent:

(1) X can be disconnected by a subset D of dimension $\leqq m$,

(2) X contains an open set U, which is neither empty nor dense, and whose boundary has dimension $\leqq m$,

(3) $X = C_1 + C_2$, with C_1, C_2 closed proper subsets of X and $\dim C_1C_2 \leqq m$.

Proof. (1) → (2). Since D disconnects X,

$$X - D = U_1 + U_2, \qquad U_1 \neq 0, \qquad U_2 \neq 0, \quad \text{and}$$

(4) $$U_1\overline{U}_2 + \overline{U}_1U_2 = 0.$$

We may assume that

(5) $$X = \overline{X - D} = \overline{U}_1 + \overline{U}_2,$$

for otherwise D would contain a non-empty open set V with $\overline{V}$ contained in D. The boundary of V, as a subset of D, would have dimension $\leqq m$, and hence would satisfy the conditions of (2).

Set $U = X - \overline{U}_1$. From (4) and (5) we conclude that $U_2 \subset U \subset \overline{U}_2$. This shows that U is not empty. Moreover, U is not dense, since $\overline{U}_2 \neq X$ by (4). Finally the boundary of U is contained in $\overline{U}_2 - U_2$, hence in D, and consequently has dimension $\leqq m$. Thus X satisfies (2).

(2) $\rightarrow$ (3). If U is a set satisfying (2), then $C_1 = \overline{U}$, $C_2 = X - U$ are sets satisfying (3).

(3) $\rightarrow$ (1). If C_1 and C_2 are sets satisfying (3) then $D = C_1C_2$ is a set satisfying (1).

Theorem IV 4. *E_n cannot be disconnected by a subset of dimension $\leqq n - 2$.*

PROOF. For if the contrary were true E_n would contain, by A), an open set which is neither empty nor dense and whose boundary has dimension $\leqq n - 2$, thus contradicting Corollary 2 to Theorem IV 3.

COROLLARY 1. *S_n, and more generally, any n-dimensional manifold, cannot be disconnected by a subset of dimension $\leqq n - 2$;*

PROOF. Suppose D were such a disconnecting set. Let the sets U_1, U_2 satisfy (4). By Theorem IV 3, D contains no open set; hence (5) holds, and because X is connected $\overline{U}_1\overline{U}_2 \neq 0$. Let p be a point of $\overline{U}_1\overline{U}_2$ and U a neighborhood of p in X which is homeomorphic to E_n. It is easy to see that U is disconnected by DU, in contradiction to Theorem IV 4.

COROLLARY 2. *I_n cannot be disconnected by a subset of dimension $\leqq n - 2$.*

PROOF. Let D be a subset of I_n of dimension $\leqq n - 2$. Let I_n' be the interior of I_n; I_n' is homeomorphic to E_n. Hence $I_n' - D$ is connected. Now all points of $I_n - D$ are cluster-points of $I_n' - D$, and therefore form a connected set also (a set remains connected upon adjunction of cluster-points).

6. Infinite-dimensional spaces

Suppose we denote by J_ω the subset of Hilbert-space consisting of points all but a finite number of whose coordinates are zero. J_ω contains a topological image of Euclidean n-space for each n; hence $\dim J_\omega = \infty$. J_ω is an example of an infinite-dimensional space which is the countable sum of finite-dimensional spaces, namely $E_1 + E_2 + E_3 + \cdots$, where E_n is the Euclidean n-space spanned by the first n coordinate axes.

J_ω is not compact, but it is also very easy to construct a compact subset K_ω of the Hilbert-cube which is the countable sum of finite-dimensional spaces: let M_n, $n = 1, 2, \cdots$ be the n-dimensional "cube" consisting of points $x = (x_1, x_2, \cdots)$ satisfying

$$|x_i| \leqq \frac{1}{i} \quad \text{for} \quad i = 1, \cdots, n,$$

$$x_i = 0 \quad \text{for} \quad i > n,$$

and let $K_\omega = \sum_1^\infty M_n$.

Not every infinite-dimensional space is the countable sum of finite-dimensional spaces; in fact

A) The Hilbert-cube is not the countable sum of finite-dimensional spaces.

Proof. By the Decomposition Theorem (Theorem III 3) this is equivalent to the statement that the Hilbert-cube is not the countable sum of zero-dimensional spaces.

Suppose the contrary:

$$(1) \qquad I_\omega = A_1 + A_2 + \cdots,$$

with each A_i of dimension zero. Let C_i be the face of I_ω determined by $x_i = 1/i$ and C_i' the face opposite. By II 2 F) there exists a closed set B_i separating C_i and C_i' and such that

$$(2) \qquad A_iB_i = 0.$$

Denote by B_i^n the intersection of B_i with the Euclidean n-space spanned by the first n coordinates of Hilbert space. Then for any given n the sets $B_1^n, B_2^n, \cdots, B_n^n$ are clearly closed sets of an interval in Euclidean n-space, each separating a different pair of opposite faces of this interval. Hence, by 1 D), these n sets have a non-zero intersection, and a fortiori, $B_1 \cdots B_n \neq 0$. The compactness of I_ω then implies that

$$(3) \qquad \prod_{i=1}^{\infty} B_i \neq 0.$$

On the other hand, (1) and (2) imply that

$$\prod_{i=1}^{\infty} B_i = 0, \tag{4}$$

and this is a contradiction.

COROLLARY. No L_p-space or l_p-space* is the countable sum of 0-dimensional spaces.

PROOF. This is an immediate consequence of the fact that all L_p- and l_p-spaces are homeomorphic† and l_2 is Hilbert space.

The possibility of decomposition into countable sums of finite-dimensional spaces is closely connected with what has been termed "transfinite-dimension." Extending Definition III 1 by transfinite induction we say, taking α as an ordinal number, that $\dim X \leqq \alpha$ if every point of X has arbitrarily small neighborhoods whose boundaries have dimension less than‡ α. We say that $\dim X = \alpha$ if $\dim X \leqq \alpha$ is true and $\dim X < \alpha$ is false.

Not every space has a transfinite dimension, but

B) If X has a transfinite dimension α then α is of the first or second ordinal class.

PROOF. We have to show that α is less than $\Omega = \omega_1$. Suppose the contrary, and let β be the smallest ordinal, $\Omega \leqq \beta \leqq \alpha$, for which there exists a space B of dimension β. By the definition of transfinite dimension B has a basis made up of open sets whose boundaries have dimension less than β, and therefore, by the minimal character of β, less than Ω. Because B is separable we may assume that this basis is countable. Now a countable collection of ordinals of the first or second ordinal-class has an upper bound in the second ordinal-class. This means that there is an ordinal

$$\gamma < \Omega$$

such that the boundary of each element of our basis has dimension less than γ. But then

$$\dim B \leqq \gamma,$$

in contradiction to

$$\dim B = \beta \geqq \Omega.$$

C) If X has a transfinite dimension then X is the countable sum of finite-dimensional subspaces.

* See index.

† S. Mazur: Une remarque sur l'homéomorphie des champs fonctionnels, *Studia Math.*, 1 (1929), pp. 83–85.

‡ $\dim B < \alpha$ means of course that $\dim B \leqq \beta$ for some $\beta < \alpha$.

Proof. Let dim $X = \alpha$. Proposition C) is obvious if $\alpha = 0$. Assume the Proposition true for all α less than β. We shall show that it is true for α equal to β. Let X be a space of dimension β. We consider a countable basis in X made up of sets having boundaries of dimension less than β.

By the hypothesis of the induction each of these boundaries is the countable sum of 0-dimensional sets. Hence the sum B of these boundaries is the countable sum of 0-dimensional sets. The proof then follows from the remark (see the proof of formula (1) on page 30) that $X - B$ is 0-dimensional.

Corollary. Hilbert-space, and the Hilbert-cube, have no transfinite dimension.

The converse of C) is not true: J_ω, the set of points in E_ω all but a finite number of whose coordinates are zero, is the countable sum of finite-dimensional spaces, but can be proved to have no transfinite dimension.

However, a partial converse does exist:

D) If X is complete* (and a fortiori, if X is compact) and the countable sum of 0-dimensional spaces, then X has a transfinite dimension.

Proof. Let

$$X = A_1 + A_2 + \cdots, \tag{5}$$

with each A_i of dimension zero. Suppose X has no transfinite dimension. Then there is a point $p \in X$ and a neighborhood U of p with this property: if V is any neighborhood of p contained in U, the boundary B of V has no transfinite dimension. Since A_1 is 0-dimensional there exists (Corollary 3 to Theorem III 2) a neighborhood V_1 of p contained in U whose boundary B_1 has no transfinite dimension, and hence is not empty, and does not meet A_1. We may also require of B_1 that its diameter is less than 1.

Now replace X by B_1 and construct in B_1 a closed set B_2 which has no transfinite dimension, does not meet A_2, and has diameter less than $\frac{1}{2}$. By induction we obtain a countable decreasing sequence of closed non-empty sets B_i whose diameters tend to zero, with

$$A_i B_i = 0. \tag{6}$$

The assumption that X is complete insures the existence of a point common to the B_i; on the other hand relations (5) and (6) make this impossible.

* See index.

CHAPTER V

Covering and Imbedding Theorems

Any subset of a Euclidean space is separable, metric, and finite-dimensional. Is the converse true?

We shall prove that it is—to be precise: every (separable metric) space of dimension less than or equal to n can be topologically imbedded in E_{2n+1} (Theorem V 3). This means that the class of finite-dimensional spaces is topologically identical with the class of subsets of Euclidean spaces.*

An example of Antonio Flores (Example V 3) shows that it is not possible to improve the number $2n + 1$, i.e. to imbed an arbitrary n-dimensional space in Euclidean space of dimension $2n$.

A further fundamental result of this chapter is the equivalence of the definition of dimension adopted in this book with the definition based on covering properties due essentially to Lebesgue (Theorem V 8). This result leads to Alexandroff's theorem on the approximation to n-dimensional spaces by n-dimensional polytopes, and connects the notion of dimension with the combinatorial properties of spaces.

1. Covering theorems

Definition V 1. By a *covering*† of a space X we mean a *finite* collection $U_1, \cdots, U_r$ of *open* sets of X whose sum is X. The *order* of a covering is the largest integer n such that there are $n + 1$ members of the covering which have a non-empty intersection. If X is bounded the *mesh* of a covering is the largest of the diameters‡ of the U_i.

Let M be a subset of X. With respect to the containing space X,

* Theorem V 3 can also be expressed in this way. If $\dim X \leqq n$, among the totality of continuous real-valued functions defined over X there exists a set of $2n + 1$ functions

$$f_1(x), \cdots, f_{2n+1}(x)$$

(the coordinate functions) which form a *basis*, in the sense that any continuous real-valued function $f(x)$ defined over X is expressible in the form

$$\varphi(f_1(x), \cdots, f_{2n+1}(x)),$$

where φ is a continuous function of $2n + 1$ real variables.

† The word "covering" is very often used in a sense more general than ours to denote an arbitrary collection of arbitrary sets whose sum is X.

‡ See index.

a finite collection of open sets of X is said to cover M if the sum of the collection contains M.

Example V 1. If each of the "bricks" in a "staggered-brick" mosaic is enlarged slightly to an open set one gets a covering of the square of order 2. It is clear that there exist coverings of this sort of arbitrarily small mesh.

Example V 2. Similarly I_n admits coverings of order n and arbitrarily small mesh.

Definition V 2. A covering β is a *refinement* of a covering α if each member of β is contained in some member of α.

A) Let X be a space and M a subset of X of dimension $\leqq 0$. Suppose U_1 and U_2 are two open sets of X which cover M. Then there exist two open sets V_1 and V_2 which cover M and satisfy

$$V_1 \subset U_1, \qquad V_2 \subset U_2, \quad \text{and}$$
$$V_1 V_2 = 0.$$

Proof. A) is evident if dim $M = -1$, i.e. if M is empty. Suppose then that M is 0-dimensional. We may assume that

$$X = U_1 + U_2,$$

for otherwise we could replace X by $U_1 + U_2$. Then

(1) $$C_1 = X - U_2, \qquad C_2 = X - U_1,$$

are closed disjoint sets. Because dim $M = 0$ we may apply II 2 F) to get a set B which is closed, does not meet M, and separates C_1 and C_2. This separation implies the existence of sets V_1 and V_2 which are open and satisfy

(2) $$V_1 \supset C_1, \qquad V_2 \supset C_2,$$

(3) $$V_1 V_2 = 0.$$

(4) $$X - B = V_1 + V_2.$$

From (1), (2), and (3) we obtain

$$V_1 \subset U_1, \qquad V_2 \subset U_2,$$

and from (4) and $MB = 0$,

$$V_1 + V_2 \supset M;$$

this proves Proposition A).

B) Let X be a space and M a subset of X of dimension $\leqq 0$ and $U_1, U_2, \cdots, U_r$ open sets covering M. Then there exist open sets $V_1, V_2, \cdots, V_r$ covering M such that

$$V_i \subset U_i, \qquad i = 1, 2, \cdots, r,$$
$$V_iV_j = 0, \ i \neq j.$$

PROOF. The proof is by induction on the number of sets U_i. Proposition B) is obvious if there is only one U_i. Now let the number of sets U_i be r and assume B) for $r - 1$. Set

$$U'_{r-1} = U_{r-1} + U_r.$$

Consider the covering of M made up of the $r - 1$ open sets

$$U_1, \cdots, U_{r-2}, U'_{r-1}.$$

By the hypothesis of the induction there exists a covering

$$V_1, \cdots, V_{r-2}, V'_{r-1}$$

of M such that

$$V_1 \subset U_1, \cdots, V_{r-2} \subset U_{r-2}, V'_{r-1} \subset U'_{r-1},$$
and $V_i, \cdots, V_{r-2}$ and V'_{r-1} are mutually disjoint.

$V'_{r-1}M$ has dimension $\leqq 0$ and U_{r-1} and U_r cover $V'_{r-1}M$. By A)' therefore, there exist open sets V_{r-1} and V_r which cover $V'_{r-1}M$ and satisfy

$$V_{r-1} \subset U_{r-1}, \qquad V_r \subset U_r,$$
$$V_{r-1}V_r = 0.$$

Then

$$V_1, \cdots, V_{r-2}, V_{r-1}, V_r$$

satisfy the condition of B).

From B) we deduce the important*

Theorem V 1. *Let X be a space of dimension $\leqq n$ and α a covering of X. Then there exists a refinement β of α of order $\leqq n$.*

PROOF. The theorem is obvious if $n = \infty$. Suppose then that n is finite. By the Decomposition Theorem (Theorem III 3)

$$X = A_1 + \cdots + A_{n+1},$$

with each summand A_i of dimension $\leqq 0$. Because α is a covering

* The converse of this theorem will be proved later (Theorem V 7).

of each A_i we may apply B) to obtain $n + 1$ finite collections β^i of open sets:

$$\beta^i = (V_1^i, \cdots, V_{r(i)}^i), \qquad i = 1, \cdots, n + 1,$$

such that

$$\beta^i \text{ is a covering of } A_i$$

and

$$V_j^i V_k^i = 0 \quad \text{if} \quad j \neq k. \tag{5}$$

Let β be the covering of X made up of all the sets V_j^i, $i = 1, \cdots, n + 1$; $j = 1, \cdots, r(i)$. Then we assert that β has order $\leqq n$. For, by the familiar argument of $n + 2$ objects in $n + 1$ boxes, any selection of $n + 2$ members of β contains two members from one of the sub-coverings β^i. Equation (5) then shows that their intersection is empty.

Corollary. *Let X be a compact metric space of dimension $\leqq n$. Then X has coverings of arbitrarily small mesh and order $\leqq n$.*

Proof. Consider, for any positive ϵ, the collection of all the spherical neighborhoods in X of radius $\frac{1}{2}\epsilon$. Since X is compact there exists a covering of X whose members are taken from this collection, i.e. a covering of X of mesh $\leqq \epsilon$. Applying Theorem V 1 to this covering, we get the Corollary.

2. Functional spaces

The considerations in Chapters V, VI, and VII, are largely based on the extremely important idea, due to Fréchet, of regarding the mappings of one space in another as themselves forming a topological space. The importance of this concept of functional space in general topology can hardly be overemphasized.

Definition V 3. Let X be an arbitrary space and Y a compact space.* We denote the set of all mappings of X in Y by Y^X, and we define a metric topology† in the *functional space* Y^X by setting

$$d(f, g) = \sup_{x \in X} d(f(x), g(x)).$$

The compactness of Y insures that $d(f, g)$ is finite. It is immediate that $d(f, g) = 0$ if and only if $f(x) \equiv g(x)$, and the triangle axiom in Y^X follows from the triangle axiom in Y.

* Whenever we talk of functional spaces in this book we consider only mappings in *compact* spaces.

† It can be shown that the topology of Y^X depends only on the topologies of X and Y and not on their metrics.

A) Y^X is complete.*

PROOF. Let f_n be a Cauchy sequence of elements of Y^X. Then for each x of X the sequence of points $f_n(x)$ in Y is a Cauchy sequence. Since Y is compact and a fortiori complete, $f_n(x)$ is convergent, to a point $f(x)$ say. We assert that $f(x)$ is continuous, i.e. $f \varepsilon Y^X$. For†

$$d(f(x), f(x')) \leqq d(f(x), f_n(x)) + d(f_n(x), f_m(x')) + d(f_m(x'), f(x')).$$

We know that f_n converges uniformly to f; hence if x' tends to x, $f(x')$ tends to $f(x)$.

B) Proposition A) enables us to apply Baire's Theorem‡ to the functional space Y^X: the intersection of a countable collection of dense G_δ-sets§ in Y^X is dense in Y^X; in particular, such an intersection is not empty.

3. Imbedding of a compact n-dimensional space in I_{2n+1}

Instead of proceeding at once to the imbedding theorem for general spaces (Theorem V 3) we shall first deal separately with compact spaces. This will make it easier to understand the main idea of the general demonstration given later, for in the case of compact spaces the proof is simpler and less obscured by technical details.

Theorem V 2. *Suppose X is a compact space and* dim $X \leqq n$, *n finite. Then X is homeomorphic to a subset of I_{2n+1}.*‖

*Moreover,¶ the set of homeomorphisms of X in*** *I_{2n+1} is a dense*†† *G_δ in the functional space $I_{2n+1}{}^X$.*

* See index.

† This is the precise analog to the proof that the uniform limit of continuous real-valued functions is continuous.

‡ See index.

§ By a G_δ-set in a space we mean any countable intersection of open sets. See Kuratowski *Topologie* I, Monografje Matematyczne, Warsaw 1933, p. 21.

‖ The statement and proof of this theorem for the case $n = 1$ are due to Menger: Allgemeine Räume und Cartesische Räume, *Proc. Akad. Wetensch. Amst.* 29 (1926), pp. 476–482; the statement for the general case is due to Menger: Über umfassendste n-dimensionale Mengen, *Proc. Akad. Wetensch. Amst.*, 29 (1926), pp. 1125–1128; the general proof, and also Theorem V 5, is due to Nöbeling: Über eine n-dimensionale Universalmenge im R_{2n+1}, *Math. Ann.* 104 (1930), pp. 71–80.

¶ For this statement, and the proof of Theorem V 2 given here, see Hurewicz: Über Abbildungen von endlichdimensionalen Räumen auf Teilmengen Cartesischer Räume, *Sitzb. Preuss. Akad. d. Wiss.*, phys. math. Klasse 1933, pp. 754–768.

** Recall that a homeomorphism "in" means a homeomorphism "on a part of."

†† The statement that the homeomorphisms form a dense set means intui-

First we consider mappings which behave "approximately" like homeomorphisms.

Definition V 4. Suppose X is a compact space. Let ϵ be a positive number and g a mapping of X in a space Y. We say that g is an *ϵ-mapping* if the inverse-image of every point of $g(X)$ has diameter less than ϵ.

A) Suppose X is a compact space. Then if g is a $1/i$-mapping of X in a space Y for each positive integer i, g is a homeomorphism of X in Y, and conversely.

Proof. If g is a $1/i$-mapping for each i, g must be one-one, and a one-one mapping of a compact space is a homeomorphism (AH, p. 95). Conversely, every homeomorphism is a $1/i$-mapping for each i.

B) Suppose X is a compact space. Then for each $\epsilon > 0$ the set G_ϵ of all ϵ-mappings is open in Y^X.

Proof. Suppose g is an ϵ-mapping. Set

$$\eta = \inf d(g(x), g(x'))$$

for $d(x, x') \geqq \epsilon$. The compactness of X insures that η is equal to $d(g(x), g(x'))$ for some pair, x, x' with $d(x, x') \geqq \epsilon$; hence η is positive, for otherwise g would not be an ϵ-mapping. Now let f be any mapping satisfying

$$d(f, g) < \tfrac{1}{2}\eta.$$

Suppose x and x' are such that $f(x) = f(x')$. It follows that $d(g(x), g(x')) < \eta$, and this implies that $d(x, x') < \epsilon$. Consequently f is also an ϵ-mapping.

Proof of Theorem V 2. Consider the functional space $I_{2n+1}{}^X$. Let $G_{1/i}$ be the set of all $1/i$-mappings (see Definition V 4) of X in I_{2n+1} and let

$$H = \prod_{i=1}^{\infty} G_{1/i}.$$

H consists of the homeomorphisms of X in I_{2n+1} by 3 A). Each $G_{1/i}$ is open by 3 B) and therefore obviously a G_δ. Hence (see 2 B)) to prove Theorem V 2 we need only the following Proposition.

C) Suppose X is a compact space and dim $X \leqq n$, n finite. For

tively that every mapping of X in I_{2n+1} can be made a homeomorphism by an arbitrarily small modification.

each positive number ϵ denote by G_ϵ the set of ϵ-mappings of X in I_{2n+1}. Then G_ϵ is dense in the functional space $I_{2n+1}{}^X$.

Proof. Let f be an arbitrary element of $I_{2n+1}{}^X$ and η a positive number. We shall construct a g such that

$$(1) \qquad d(f, g) < \eta,$$

$$(2) \qquad g \in G_\epsilon.$$

The uniform continuity* of f yields a positive number $\delta < \epsilon$ such that

$$d(f(x), f(x')) < \tfrac{1}{2}\eta$$

whenever

$$d(x, x') < \delta.$$

By the Corollary to Theorem V 1, there exists a covering

$$\beta\colon \quad U_1, \cdots, U_r$$

of X with the properties

$$(3) \qquad \text{order} \quad \beta \leqq n$$

and†

$$(4) \qquad \delta(U_i) < \delta, \qquad i = 1, \cdots, r.$$

In consequence,

$$(5) \qquad \delta(f(U_i)) < \tfrac{1}{2}\eta, \qquad i = 1, \cdots, r.$$

Select vertices $p_1, \cdots, p_r$ in I_{2n+1} for which it is true that

$$(6) \qquad d(p_i, f(U_i)) < \tfrac{1}{2}\eta, \qquad i = 1, \cdots, r,$$

(7) the p_i are in general position in E_{2n+1},

i.e. no $m + 2$ of the vertices p_i ($m = 0, 1, \cdots, 2n$) lie in an m-dimensional linear subspace of E_{2n+1}.

For each point x of X set‡

$$w_i(x) = d(x, X - U_i), \qquad i = 1, \cdots, r.$$

* A continuous function on a compact space is uniformly continuous.

† $\delta(M)$ denotes the diameter of M; see index.

‡ If $U_i = X$ we agree that $w_i(x) = 1$.

Evidently

$$w_i(x) > 0 \quad \text{if} \quad x \in U_i \quad \text{and} \quad w_i(x) = 0 \quad \text{if} \quad x \notin U_i.$$

For each x at least one $w_i(x)$ is positive, since the U_i cover X. Assign to each of the vertices p_i chosen in the paragraph above the weight $w_i(x)$, and denote by $g(x)$ the center of gravity of the system of vertices p_i with these weights. It is clear that this transformation* g of X in I_{2n+1} is continuous. We now show that g satisfies (1) and (2).

To prove (1): Let x be an arbitrary point of X and assume the U_i are so numbered that $U_1, \cdots, U_s$ is the set of all the U_i which contain x. Then $w_i(x) > 0$ for $i \leqq s$ and $w_i(x) = 0$ for $i > s$; hence in the definition of $g(x)$ we need only consider $p_1, \cdots, p_s$. From $x \in U_i$, $i \leqq s$, and (5) and (6), we get

$$d(p_i, f(x)) < \eta, \qquad i \leqq s.$$

A fortiori, the center of gravity $g(x)$ of the p_i satisfies

$$(1) \qquad d(g(x), f(x)) < \eta.$$

To prove (2): Suppose

$$U_{i_1}, \cdots, U_{i_s}$$

are all the members of β containing a given point x of X. (3) then implies that $s \leqq n + 1$. Consider the linear $(s-1)$-space $L(x)$ in I_{2n+1} spanned by the vertices

$$p_{i_1}, \cdots, p_{i_s}.$$

It is clear that $g(x)$ is in $L(x)$. Let x' be another point of X. We assert: *if $L(x)$ and $L(x')$ meet they contain a common vertex p_i; hence x and x' are contained in a common member U_i of β.* For suppose $L(x')$ is spanned by

$$p_{j_1}, \cdots, p_{j_t}.$$

Again by (3), $t \leqq n + 1$ and $L(x')$ is a $(t-1)$-space. If $L(x)$ and $L(x')$ meet, the linear space spanned by the p_i and the p_j together has dimension $\leqq s + t - 2 \leqq 2n$. From this and (7) it follows that at least one of the p_i is also a p_j, which establishes the assertion.

Now suppose $g(x) = g(x')$. Then $L(x)$ and $L(x')$ meet. By the assertion above, x and x' are contained in a common member of β. (4) then shows that $d(x, x') < \delta < \epsilon$, and consequently g is an

* Such "barycentric" mappings have many important applications, and will be discussed in detail in Section 9.

ϵ-mapping. This completes the proof of Proposition C) and hence that of Theorem V 2.

4. Imbedding of an n-dimensional space in I_{2n+1}

Using the proof of Theorem V 2 as a model we now prove

Theorem V 3. *Suppose X is an arbitrary space and* dim $X \leqq n$, *n finite. Then X is homeomorphic* to a subset of I_{2n+1}.*

Moreover† the set of homeomorphisms of X in I_{2n+1} contains‡ a dense G_δ in the functional space $I_{2n+1}{}^X$.

The ϵ-mappings used in Section 3 are inadequate for our present purpose, because a mapping of an arbitrary space which is an ϵ-mapping for each $\epsilon > 0$ (and hence one-one) need not be topological: this is illustrated by the mapping

$$\theta = x$$

of the half-open line interval $0 \leqq x < 2\pi$ on the circumference $0 \leqq \theta \leqq 2\pi$.

Now in the general question of the transition from compact spaces to arbitrary spaces the following principle has proved to be of great value: replace "for every positive ϵ there is a set of diameter less than $\epsilon \cdots$" by "for every covering α there is a refinement of $\alpha \cdots$." Using this principle we are led to this modification of Definition V 4:

Definition V 5. Let α be a covering of X and g a mapping of X in a space Y. We say that g is an α-mapping if every point of Y has a neighborhood in Y whose inverse-image is entirely contained in some member of α.

Definition V 6. Let α be a covering of X. Denote by $S_\alpha(x)$ the open set which is the sum of the members of α containing a given point x. A countable collection $\alpha^1, \alpha^2, \cdots$ of coverings is called a

* This result was first obtained by combining the Menger-Nöbeling theorem (the first part of Theorem V 2) with Hurewicz's proof (Über das Verhältnis separabler Räume zu kompakten Räumen, *Proc. Akad. Wetensch. Amst.* 30 (1927), pp. 425–430) that any space is topologically contained in a compact space of the same dimension (in the present book this statement appears (Theorem V 7) as a consequence of the general imbedding theorem).

† First proved by Kuratowski (Sur les théorèmes de "plongement" dans la théorie de la dimension, *Fund. Math.* 28 (1937), pp. 336–342), whose method is a slight modification of that of Hurewicz in his Sitzgb. paper cited on page 56.

‡ Note the difference between this statement and the corresponding statement in the compact case. It has been announced by J. H. Roberts (Abstract 160, *Bull. Am. Math. Soc.* 53 (1947), p. 287) that the set of homeomorphisms of X in I_{2n+1} is not necessarily a G_δ.

basic sequence of coverings if given a point x and a neighborhood U of x at least one of the open sets

$$S_{\alpha^1}(x),\ S_{\alpha^2}(x),\ \cdots,$$

is contained in U.

A) For each space X there exists a basic sequence of coverings.

Proof. Let $U_1, U_2, \cdots$ be any countable basis of X. Consider pairs U_n, U_m of these basic non-empty open sets for which

$$\overline{U}_n \subset U_m.$$

Denote by $\alpha^{n,m}$ the covering of X whose two members are $X - \overline{U}_n$ and U_m. The collection of the coverings $\alpha^{n,m}$ is of course countable. Moreover, $x \in U_n$ implies $S_{\alpha^{n,m}}(x) = U_m$. Hence the collection of sets $\{S_{\alpha^{n,m}}(x)\}$ for a given x includes the collection of all the U_m containing x, and this proves that $\{\alpha^{n,m}\}$ is a basic family.

B) Suppose $\alpha^1, \alpha^2, \cdots$ is a basic sequence of coverings of a space X. Then if g is an α^i-mapping of X in a space Y for every i, g is a homeomorphism.*

Proof. We shall demonstrate that if x is any point of X and U a neighborhood of x in X there is a neighborhood V of $g(x)$ in Y whose inverse image is contained in U. From this follows the one-one character of g and the continuity of g^{-1}.

By the definition of a basic sequence of coverings there is an α^i for which

(1) $$S_{\alpha^i}(x) \subset U.$$

Since g is an α^i-mapping there is a neighborhood V of $g(x)$ and a member U_0^i of α^i for which

(2) $$\overset{-1}{g}(V) \subset U_0^i.$$

But

$$x \in \overset{-1}{g}(V) \subset U_0^i;$$

* The converse is not true. For let X be the Euclidean half-line $0 \leqq x$, Y the segment $0 \leqq y \leqq 1$, and h the "shrinking" of X in Y given by

$$y = \frac{x}{1 + x}.$$

Let α^0 be the covering of X whose two members are the complements of the even and odd integers respectively. Although h is a homeomorphism, h is not an α^0-mapping, since no neighborhood of the point $y = 1$ has its inverse-image entirely contained in a member of α^0.

hence

(3) $$U_0^i \subset S_{\alpha^i}(x)$$

(2), (3), and (1) prove the proposition: $g^{-1}(V) \subset U$.

C) Suppose X is a compact space and α is a covering of X. Then there is a positive number η with the property that every subset of X of diameter less than η is entirely contained in some member of α.

PROOF. Otherwise there would be a sequence $X_1, X_2, \cdots$ of subsets with diameters tending to zero and not contained in any member of α. Let $x_i \in X_i$. Since X is compact, $\{x_i\}$ has a cluster-point x, which is contained in some member, U_0 say, of α. But U_0 is open, so that x has a positive distance d from $X - U_0$. Then every X_i of diameter less than d is contained in U_0, contrary to the assumption.

D) Let X be any space and Y a compact space. For each covering α of X the set G_α of all α-mappings of X in Y is open in Y^X.

PROOF. Suppose g is an α-mapping. This means that each point of Y has a neighborhood whose inverse-image is entirely contained in some member of α. Since Y is compact there is a finite sub-collection of these neighborhoods which form a covering σ of Y. One derives from C) a positive number η with the property that any set in Y of diameter less than η is contained in a member of σ and hence has its inverse-image under g entirely contained in some member of α. Now let f be any mapping satisfying

$$d(f, g) < \tfrac{1}{3}\eta.$$

Take the spherical neighborhoods of diameter $\frac{1}{3}\eta$ around each point of Y. Let A be the inverse-image under f of one of these neighborhoods. It is easy to see that $g(A)$ has diameter $< \eta$, so that by the definition of η, A is contained in some member of α. This shows that f is an α-mapping.

PROOF OF THEOREM V 3. Consider the functional space $I_{2n+1}{}^X$. Let $\alpha^1, \alpha^2, \cdots$ be a basic-sequence of coverings of X, G_{α^i} the set of all α^i-mappings (see Definition V 5) of X in I_{2n+1}, and let

$$H = \prod_{i=1}^{\infty} G_{\alpha^i}.$$

Each element of H is a homeomorphism of X in I_{2n+1} by 4 B). Each G_{α^i} is open by 4 D), and therefore obviously a G_δ-set. Hence (see 2 B)) to prove Theorem V 3 we need only the following Proposition.*

E) Suppose dim $X \leqq n$, n finite. For each covering α of X denote by G_α the set of α-mappings of X in I_{2n+1}. Then G_α is dense in the functional space $I_{2n+1}{}^X$.

PROOF. Let f be an arbitrary element of $I_{2n+1}{}^X$ and η a positive number. We shall construct a g such that

$$d(f, g) < \eta, \tag{4}$$

$$g \in G_\alpha. \tag{5}$$

As a compact space, I_{2n+1} has a covering of mesh less than $\frac{1}{2}\eta$; let τ be the covering of X made up of the inverse-images of these open sets under the continuous function f. By Theorem V 1 there exists a common refinement β: $U_1, \cdots, U_r$ of α and τ of order $\leqq n$:

$$\text{order } \beta \leqq n, \tag{6}$$

$$\beta \text{ a refinement of } \alpha, \tag{7}$$

$$\delta(f(U_i)) < \tfrac{1}{2}\eta, \qquad i = 1, \cdots, r. \tag{8}$$

We now construct, g exactly as in the proof of 3c) above, as a "barycentric"-mapping based on r points $p_1, \cdots, p_r$ in I_{2n+1} in general position. The proof of (4) is exactly as before, and in the proof of (5) we need only replace the last paragraph by the following:

Since there are only a finite number of the linear subspaces $L(x)$ there exists a number $\eta > 0$ such that any two of these linear subspaces $L(x)$ and $L(x')$ either meet or else have a distance $\geqq \eta$ from each other. If $d(g(x), g(x')) < \eta$ the distance $d(L(x), L(x'))$ is certainly $< \eta$; hence $L(x)$ meets $L(x')$. As shown above this implies that x and x' are contained in a common member of β. Consequently g is an α-mapping.

EXAMPLE V 3. Let s_{2n+2} be a $(2n + 2)$-dimensional cell† and P_n the collection of all faces of s_{2n+2} of dimension $\leqq n$. Then P_n is an n-dimensional space which cannot be imbedded in E_{2n}. The proof will be found in the paper of Flores "Über n-dimensionale Komplexe die im R_{2n+1} absolut selbstverschlungen sind," *Ergebnisse eines mathematischen Kolloquiums* 6 (1933–4), pp. 4–7, and shows that the number $2n + 1$ in Theorem V 3 is best possible.

* Compare this with the analogous part (Proposition 3C)) of the proof of Theorem V 2.

† See page 67.

5. Imbedding of arbitrary spaces in the Hilbert cube

Theorem V 4. *Let X be any space. Then X can be topologically imbedded* in I_ω. Moreover the set of homeomorphisms of X in I_ω contains a dense G_δ-set in $I_\omega{}^X$.*

PROOF. The proof is almost exactly the same as that of Theorem V 3, with the simplification that we are not concerned with the order of the covering β. The points p_i are chosen so that any finite subset of them are linearly independent.

6. A universal n-dimensional space

Definition V 7. An n-dimensional space is a *universal n-dimensional space* if every space of dimension $\leqq n$ can be topologically imbedded in it.

Theorem V 5.† *The set*

$$\mathcal{X}_n = \overset{n}{\mathcal{M}}_{2n+1} \cdot I_{2n+1}$$

of points in I_{2n+1} at most n of whose coordinates are rational is a universal n-dimensional space.

PROOF. We have to show that any space of dimension $\leqq n$ can be topologically imbedded in $\mathcal{X}_n$. The proof is a modification of that of Theorem V 3 and uses the same notation. It is easy to see that if M is a fixed n-dimensional linear subspace of E_{2n+1}, one can slightly move the vertices $p_1, \cdots, p_r$ so that none of the $L(x)$ (see 3 C) and 4 E)) meets M. Hence starting with an arbitrary mapping f of X in I_{2n+1} and positive number η we can construct a mapping g satisfying $d(f, g) < \eta$ and in addition the condition:

$$\overline{g(X)} \subset I_{2n+1} - M. \tag{1}$$

It follows that the set of mappings g of X in I_{2n+1} for which (1) holds

* This is a celebrated theorem of Urysohn: Zum Metrisationsproblem, *Math. Ann.* 94 (1925), pp. 309–315. X is, of course, separable metric.

† Nöbeling: Über eine n-dimensionale Universalmenge im R_{2n+1}, *Math. Ann.* 104 (1930), pp. 71–80.

Another universal n-space was described by Menger, but without proof: Über umfassendste n-dimensionale Mengen, *Proc. Akad. Wetensch. Amst.* 29 (1926), pp. 1125–1128; the proof is due to Lefschetz: On compact spaces, *Ann. Math.* 32 (1931), pp. 521–538.—Menger's universal space has the additional property of being compact. For $n = 0$ Nöbeling's and Menger's universal spaces specialize to the set of irrationals in the line segment and the Cantor set, respectively.

Observe that the existence of a *compact* universal n-space follows from application of Theorem V 6 below to the universal n-space $\mathcal{X}_n$.

is dense in $I_{2n+1}{}^X$. This set is also open, for $\overline{g(X)} \subset I_{2n+1} - M$ implies that $g(X)$ has a positive distance η from M, and then every mapping f with $d(f, g) < \eta$ satisfies $\overline{f(X)} \subset I_{2n+1} - M$.

Now the complement of $\mathfrak{X}_n$ consists of the points of I_{2n+1} at least $n + 1$ of whose coordinates are rational, i.e. the complement of $\mathfrak{X}_n$ is the sum of those hyperplanes in I_{2n+1} of the form

$$x_{i_1} = r_1, \cdots, x_{i_{n+1}} = r_{n+1},$$

the r's being rational. Each of these hyperplanes has dimension $2n + 1 - (n + 1) = n$, and there are a countable number of them. Call them $M_1, M_2, \cdots$. Let G_i be the open dense subset of $I_{2n+1}{}^X$ consisting of all mappings g for which

$$\overline{g(X)} \subset I_{2n+1} - M_i.$$

From Theorem V 3 there is a dense G_δ-set H contained in $I_{2n+1}{}^X$ whose members are homeomorphisms of X in I_{2n+1}. Let

$$H' = H \prod_{i=1}^{\infty} G_i.$$

H', as a countable intersection of dense G_δ-sets, is itself a dense G_δ (see 2 B)); in particular, H' is not empty. Let $h \in H'$. It is clear that h is a homeomorphism of X in I_{2n+1}, and

$$\overline{h(X)} \subset I_{2n+1} - (M_1 + M_2 + \cdots) = \mathfrak{X}_n. \tag{2}$$

Hence for every space of dimension $\leqq n$ there exists a homeomorphism h of X in I_{2n+1} such that

$$\overline{h(X)} \subset \mathfrak{X}_n. \tag{3}$$

But $\mathfrak{X}_n$ has dimension n (see Example IV 1). Hence the theorem is proved.

Remark. It is not known where $\mathfrak{M}_n^k$, $n < 2k + 1$ contains a topological image of every k-dimensional subset of E_n, or even whether there exists any* k-dimensional subset of E_n containing a topological image of every k-dimensional subset of E_n.

Theorem V 6. *Any space can be topologically imbedded in a compact space of the same dimension.*

Proof. Let X be the space. The theorem is obvious if $\dim X = \infty$,

* See Menger: Über umfassendste n-dimensionale Mengen, *Kon. Akad. v. Wetensch. Amst.* 29 (1926), pp. 1125–1128.

for any space can be topologically imbedded in I_ω (Theorem V 4). If dim $X \leqq n$, n finite, the theorem follows from formula (3) in the proof of Theorem V 5 above, since $\overline{h(X)}$, as a closed subset of I_{2n+1}, is compact.

7. Fréchet dimension-type

By analogy with the theory of cardinal numbers, Fréchet introduced in 1909 the concept of dimension-type, saying of two given spaces A and B that

$$\text{dimension-type } A \leqq \text{dimension-type } B$$

if A can be topologically imbedded in B. If it is also true that

$$\text{dimension-type } B \leqq \text{dimension-type } A$$

one says that A and B have the same dimension-type. Euclidean n-space is said to have dimension-type n.

It is obvious that two spaces of the same dimension-type have the same dimension. The simple example of a circumference and an arc shows, on the other hand, that two spaces of the same dimension need not have the same dimension type.

Theorem V 3 may be rephrased thus: if dim $X \leqq n$ then dimension-type $X \leqq 2n + 1$. Theorem V 5 states that among the dimension-types of spaces of dimension $\leqq n$ there is a greatest dimension-type. Theorem V 4 asserts that every space has dimension-type less than or equal to that of the Hilbert-cube.

The discussion of Section 6 of Chapter IV shows that there are at least two dimension-types among infinite-dimensional spaces, namely those of the Hilbert-cube itself, and of a subset of the Hilbert-cube which is a countable sum of finite-dimensional spaces.

8. Covering theorems again

Theorem V 7. *If every covering of a space X has a refinement of order $\leqq n$ then X has dimension $\leqq n$.*

PROOF. For in the proof of Theorem V 5 we used only this covering property of X to deduce that X could be imbedded in the universal n-dimensional space $\mathfrak{X}_n$.

COROLLARY. *Suppose X is a compact space. If X has coverings of arbitrarily small mesh and order $\leqq n$ then X has dimension $\leqq n$.*

PROOF. From 4 C) one concludes that every covering of X has a refinement of order $\leqq n$. The corollary is then immediate.

Theorem V 8. Covering Theorem. *A space has dimension $\leqq n$ if and only if every covering has a refinement of order $\leqq n$.*

PROOF. This is a combination of Theorems V 1 and V 7.

COROLLARY. **Covering Theorem for Compact Spaces.** *A compact space has dimension $\leqq n$ if and only if it has coverings of arbitrarily small mesh and order $\leqq n$.*

PROOF. This is a combination of the Corollary to Theorem V 1 and the Corollary to Theorem V 7.

9. Nerves and mappings in polytopes

We shall consider polytopes in the most unsophisticated sense, as rectilinear point sets in a Euclidean space. A *vertex* or *0-cell* is a point; a *1-cell* a segment without its end points; a *2-cell* a triangle without its sides; a *3-cell* a tetrahedron without its faces; and so on. The k-cells, $k = 0, 1, 2, \cdots$, determined by the vertices, sides, faces, $\cdots$ of a p-cell are called the *k-faces* of the p-cell; we also include the p-cell itself among its faces. An n-polytope is a point-set contained in some E_m and decomposed in a definite manner in a finite collection of disjoint p-cells, $0 \leqq p \leqq n$, at least one of which is an n-cell, and such that every face of each cell of the collection belongs to the collection.

Now consider any set of objects, which we shall call *(abstract) vertices.* By an *(abstract) p-simplex*, s_p, $p = 0, 1, 2, \cdots$ we mean any set of $p + 1$ vertices. A k-simplex whose vertices are chosen from those of s_p is called a *k-face* of s_p; s_p is a face, the p-face, of itself. An *(abstract) n-complex* is a finite collection of p-simplexes, $0 \leqq p \leqq n$, which contains every face of each simplex of the collection and at least one n-simplex.

The connection between polytopes and complexes is brought into view by the following two statements:

A) With each polytope P we associate the complex N, called the *vertex-scheme* of P, whose vertices are identified with those of P and whose simplexes are those collections of vertices which span a cell of P.

Conversely,

B) Given an n-complex N there is an n-polytope P, called a *geometrical realization* of N, whose vertex-scheme is N; moreover, P may be taken as a subset of I_{2n+1}.

PROOF. Let $p_1, \cdots, p_r$ be all the vertices of N. It can easily be proved that we can select r vertices in I_{2n+1}, which we continue to call

$p_1, \cdots, p_r$, in general position, i.e. any $m + 2$, $m \leqq 2n$, of these points are linearly independent. We now let P be the collection of all cells in I_{2n+1} spanned by vertices $p_{i_0}, \cdots, p_{i_k}$ for which $(p_{i_0}, \cdots, p_{i_k})$ is a simplex of N. If P were a polytope its vertex-scheme would be N, so that it remains only to show that P is a polytope. To do this we must prove that any two cells s and t of P are disjoint. Let $p_1, \cdots, p_k$ be all the distinct vertices of s and t. Each of the cells of P has dimension $\leqq n$; hence $k \leqq 2n + 2$. Because the points $p_1, \cdots, p_r$ are in general position, the points $p_1, \cdots, p_k$ are linearly independent and therefore span a $(k - 1)$-cell u (not necessarily in P) which has s and t among its faces. Our assertion then follows from the fact that any two distinct faces of a cell are disjoint.

Remark. Two polytopes with the same vertex-scheme can easily be proved to be homeomorphic. Hence we may speak of *the* geometrical realization of a complex.

Alexandroff* has introduced the following very useful process associating with each covering of a space a complex called its nerve. The concept of the nerve of a covering is of great importance in modern topology, for it forms a link between the continuous and the combinatorial methods. Nerves may be regarded as combinatorial configurations approximating to the space; the finer the covering the better the approximation.

Definition V 8. Let α: $U_1, \cdots, U_r$ be a covering of a space. We associate with each non-empty U_i a mark p_i, and with the p_i as vertices we construct a complex called the *nerve* $N(\alpha)$ of α in this way:

$$(p_{i_1}, \cdots, p_{i_k})$$

is a simplex of $N(\alpha)$ if and only if

$$U_{i_1} \cdots U_{i_k} \neq 0.$$

It is clear that

$$\dim N(\alpha) = \text{order } \alpha.$$

We denote by $P(\alpha)$ the geometrical realization of $N(\alpha)$.

Closely connected with the concept of nerve is that of "barycentric" mapping:

* Über den allgemeinen Dimensionsbegriff und seine Beziehungen zur elementaren geometrischen Anschauung, *Math. Ann.* 98 (1928), pp. 617–635, in particular p. 634.

Definition V 9. Let X be a space and α a covering of X. Let P be a polytope whose vertices* $p_1, \cdots, p_r$ are in one-one correspondence with the members U_i of α. Let Z_i be the *star* of p_i, i.e. the open set in P composed of cells having p_i as vertex. A mapping g of X in P is called a *barycentric α-mapping* if

$$g^{-1}(Z_i) = U_i.$$

C) Since the Z_i form a covering of P, a barycentric α-mapping is an α-mapping (see Definition V 5).

D) Suppose g is a barycentric α-mapping of X in a polytope P and P' is the subpolytope of P consisting of all faces of cells containing a point of $g(X)$. Then the vertex-scheme of P' is precisely $N(\alpha)$, or what is the same, P' is the geometrical realization of $N(\alpha)$.

Proof. First we show that if s' is any cell of P', which we denote by $(p_1, \cdots, p_m)$, then

$$\prod_{i=1}^{m} U_i \neq 0, \tag{1}$$

i.e. s' is also a simplex of $N(\alpha)$. By the definition of P' there is a cell s, denoted by $(p_1, \cdots, p_m, p_{m+1}, \cdots, p_k)$, containing a point $y_0 \, \varepsilon \, g(X)$ and having s' as face. Now

$$y_0 \, \varepsilon \, Z_i, \qquad i = 1, \cdots, k.$$

Hence

$$g^{-1}(y_0) \subset g^{-1}(Z_i) = U_i, \qquad i = 1, \cdots, k,$$

and this proves (1).

Conversely, let s: $(p_1, \cdots, p_m)$ be any simplex of $N(\alpha)$. Then $U_1, \cdots, U_m$ have a point x_0 in common.

$$g(x_0) \, \varepsilon \, Z_i, \qquad i = 1, \cdots, m,$$

and since $g(x_0)$ is a point of P', one concludes that

$$g(x_0) \, \varepsilon \, Z_i', \qquad i = 1, \cdots, m,$$

where Z_i' denotes the star, in P', of p_i. Hence the stars Z_i', $i = 1, \cdots, m$, have a non-zero intersection, which implies that $(p_1, \cdots, p_m)$ is a cell of P'. This completes the proof of D).†

* The p_i will also be taken as the vertices of $N(\alpha)$.

† An interesting special case of D) is the one in which $X = P$, α is the covering of P made up of the stars of the vertices of P, and g is the identity mapping of P on P. In this case P' is, of course, the same as P so that D) asserts that *the nerve of α is the vertex scheme of P.*

Proposition D) shows that without loss of generality we may restrict the study of barycentric α-mappings to the case that P is the geometrical realization of the nerve of α.

The "barycentric" in "barycentric α-mapping" is justified by the following Proposition.

E) Barycentric α-mappings of X in $P(\alpha)$ coincide with those mappings g obtained by this construction: Let $U_1, \cdots, U_r$ be the members of α. Define r continuous real-valued functions $w_i(x)$ such that

$$w_i(x) = 0 \quad \text{if} \quad x \not\varepsilon U_i, \tag{2}$$

$$w_i(x) > 0 \quad \text{if} \quad x \,\varepsilon\, U_i. \tag{3}$$

$g(x)$ is the correspondence assigning to each $x \,\varepsilon\, X$ the center of gravity of the vertices p_i with the weights $w_i(x)$.

Proof. Suppose g is a barycentric α-mapping. Given $x \,\varepsilon\, X$, we define $w_i(x)$ as follows: Let $s = (p_{i_1}, \cdots, p_{i_k})$ be the cell of $P(\alpha)$ containing $g(x)$. Let

$$\begin{aligned} w_i(x) &= 0 \text{ if } i \text{ is different from each of } i_1, \cdots, i_k \\ w_i(x) &= \text{the barycentric coordinate* of } x \text{ with respect to} \\ &\quad\; p_i \text{ if } i \text{ is one of } i_1, \cdots, i_k. \end{aligned}$$

$w_i(x) > 0$ if and only if $p_i \,\varepsilon\, s$, i.e. $g(x) \,\varepsilon\, Z_i$. Hence the functions $w_i(x)$ satisfy (2) and (3).

The converse is proved similarly.

F) Functions $w_i(x)$ satisfying (2) and (3) of E) can always be found,† i.e. *for each covering α of a space X there is a barycentric α-mapping of X in $P(\alpha)$.*

Proof. For example:

$$w_i(x) = d(x, X - U_i).$$

(If $X - U_i$ is empty we set $w_i(x) = 1$.)

Remark 1. If we examine the proofs of Propositions 3 C) and 4 E) we see that the mapping g constructed there is such that $\overline{g(X)}$ is contained in an n-dimensional polytope contained in I_{2n+1}; g is, in fact, a barycentric mapping of X in the geometric realization of the nerve of

* The barycentric coordinates of a point p of a cell $(p_1, \cdots, p_k)$ are the weights $w_1, \cdots, w_k$, of sum 1, which must be assigned to the respective vertices in order to get p as center of gravity.

† Proposition F) holds, somewhat more generally, for "perfectly normal" spaces, even if they are not metrizable.

the covering β. This remark permits the following strengthening of 4 E):

4 E′) Suppose dim $X \leqq n$, n finite. For each covering of X denote by G_α' the set of α-mappings of X such that $\overline{g(X)}$ is contained in an n-dimensional polytope contained in I_{2n+1}. Then G_α' is dense in the functional space $I_{2n+1}{}^X$.

Remark 2. The reader will now see that the proof of the imbedding theorem (Theorem V 3) involves the following ideas:

(a) an n-dimensional space has arbitrarily fine coverings of order n (Theorem V 1);

(b) the nerve of a covering of order n can be geometrically realized by a polytope in I_{2n+1} (Proposition B));

(c) a space can be barycentrically mapped on the geometrical realization of any of its nerves (Proposition F)).

(d) Baire's Theorem.

We now prove

Theorem V 9. Approximation by Polytopes. *A space X has dimension $\leqq n$ if and only if for every covering α of X there is an α-mapping of X in a polytope of dimension $\leqq n$.*

Proof. Necessity. Suppose dim $X \leqq n$. By Theorem V 1 there exists a refinement α' of α of order $\leqq n$. By F) there is an α'-mapping (in fact, a barycentric α'-mapping) of X in $P(\alpha')$. But an α'-mapping is a fortiori an α-mapping, and $P(\alpha')$ has dimension $\leqq n$.

The proof of sufficiency is contained in the following Proposition:

G) Suppose dim $X \geqq m$. Then there exists a covering α of X with this property: for every α-mapping g of X in a compact space Y,

$$\dim g(X) \geqq m.$$

Proof. By Theorem V 7 there is a covering α of X each of whose refinements has order $\geqq m$. Let g be an α-mapping of X in Y. Each point of Y has a neighborhood whose inverse-image is contained in an element of α. Since Y is a compact a finite number of these neighborhoods cover Y, and the intersections with $g(X)$ of this finite number of neighborhoods form a covering β of $g(X)$. Suppose dim $g(X) < m$. Then by Theorem V 1, β has a refinement β' of order $< m$. The inverse-images of the elements of β' would form a refinement of α of order $< m$, contrary to the hypothesis on α. Hence G) is proved, and with it Theorem V 9.

Corollary. **Alexandroff's Theorem on Approximation to Compact Spaces by Polytopes.*** *Suppose X is a compact space. Then* dim $X \leqq n$ *if and only if for every positive ϵ there is an ϵ-mapping of X in a polytope of dimension $\leqq n$.*

Proof. If dim $X \leqq n$ it follows from Theorem V 9 that for each covering α there is an α-mapping g_α of X in a polytope of dimension $\leqq n$. Given $\epsilon > 0$, let α be a covering of X of mesh $< \epsilon$. Such a covering exists by the compactness of X. For this α, g_α is an ϵ-mapping.

Conversely, suppose there is an ϵ-mapping g_ϵ of X in a polytope of dimension $\leqq n$ for each positive ϵ. Given an arbitrary covering α, let ϵ be the positive real number (see 4 C)) such that any set of diameter less than ϵ is contained in one of the members of α. For this ϵ, g_ϵ is an α-mapping. Hence dim $X \leqq n$, by Theorem V 9. This proves the corollary.

Let X be a space, α a covering of X and $P = P(\alpha)$ the geometrical realization of the nerve of α. A mapping f of X in $P(\alpha)$ is called a *quasi*-barycentric α-mapping† if (cf. Definition V 9)

$$f^{-1}(Z_i) \subset U_i.$$

* This theorem has been basic for many recent topological researches. See Alexandroff, "Über den allgemeinen Dimensionsbegriff und seine Beziehungen zur elementaren geometrischen Anschauung," *Math. Ann.* 98 (1928), 617–635.

† It is not hard to show that quasi-barycentric α-mappings coincide with those determined by continuous real-valued functions $w_i(x)$ for which

(2′) $$w_i(x) = 0 \quad \text{if} \quad x \notin U_i,$$

(3′) $$w_i(x) \geqq 0 \quad \text{if} \quad x \in U_i,$$

(4′) $$\sum w_i(x) > 0 \quad \text{for every } x.$$

(Compare these formulas with (2) and (3) of E).) Given any normal space X, separable or not, and any covering α: $U_i, \cdots, U_r$ of X, there is always a quasi-barycentric α-mapping of X in $P(\alpha)$. This is proved as follows: The normality of X yields for each U_i an open set V_i such that:

$$\overline{V}_i \subset U_i,$$

the V_i form a covering α' of X,

the nerves of α and α' are identical.

Normality also yields r continuous real valued functions $w_i(x)$ for which

$$0 \leqq w_i(x) \leqq 1,$$

$$w_i(x) = 0 \quad \text{for every} \quad x \notin U_i,$$

$$w_i(x) = 1 \quad \text{for every} \quad x \in \overline{V}_i.$$

These functions $w_i(x)$ thus satisfy (2′), (3′), and (4′) above.

A quasi-barycentric α-mapping is, of course, an α-mapping.

H) Consider all the subpolytopes P' of $P(\alpha)$ which have this property:

(4) there is a quasi-barycentric α-mapping f of X in P such that $f(X) \subset P'$.

Let P_0 be a subpolytope of P which is irreducible with respect to this property, i.e. P_0 satisfies (4) while no proper subpolytope of P_0 does. Such a polytope obviously exists. Let f_0 be a quasi-barycentric α-mapping of X in $P(\alpha)$ for which $f_0(X) \subset P_0$. Then

$$f_0(X) = P_0.$$

Proof. Suppose $p_0 \varepsilon P_0$ were not contained in $f_0(X)$. Let s be the cell of P_0 containing p_0, and let S be the *star of the cell* s, i.e. the set of all cells having s as face. We denote by $B(S)$ (the "boundary" of S) the set of all cells which are not in S but are faces of cells of S. For each $x \varepsilon X$ whose image $f_0(x)$ is contained in S we replace $f_0(x)$ by its projection $f'(x)$ on $B(S)$ from p_0. One easily proves that $f'(x)$ is also a quasi-barycentric α-mapping of X in P. But $f'(X)$ is contained in a proper subpolytope of P_0, contradicting the definition of P_0.

This shows that in Theorem V 9 mapping "in" can be replaced by mapping "on":

Theorem V 10.* *A space X has dimension $\leqq n$ if and only if for every covering α of X there is an α-mapping of X* **on** *a polytope of dimension $\leqq n$.*

Proof. Only the necessity need be proved. Suppose dim $X \leqq n$. Replace α by a refinement α' of order $\leqq n$, and let f_0 be the α'-mapping of X, whose existence is a consequence of H), *on* the irreducible polytope P_0; f_0 is the required mapping.

* This theorem, in the most important case of compact spaces, is due to Alexandroff (see footnote* on page 72).

CHAPTER VI

Mappings in Spheres and Applications

In this chapter we study mappings of topological spaces in spheres, and characterize dimension in terms of such mappings. This characterization, contained in Theorem VI 4, is the main result of the chapter and will find its most important application in Chapter VIII, where it serves as the basis for the algebraic and combinatorial treatment of dimension theory.

The technique of mappings in spheres and the all-important notion of homotopy are further applied to obtain simple proofs of separation properties in Euclidean spaces (n-dimensional Jordan theorem, §7), and to investigate the changes in dimension effected by continuous transformation of a space, §4.

1. Stable and unstable values

In this section we investigate the following problem: Under what conditions is it possible to map a space X in the Euclidean n-space so that at least one point of E_n is covered "essentially," i.e. cannot be uncovered by arbitrarily small modifications of the mapping? In other words: Under what conditions is it possible to define n continuous real-valued functions

$$f_i(x), \qquad x \in X, \qquad i = 1, \cdots, n$$

in such a way that for any system of continuous functions $g_i(x)$ approximating the functions $f_i(x)$ sufficiently closely the system of equations

$$g_i(x) = 0$$

has a solution in x?

It turns out (Theorems VI 1 and VI 2) that such a system of functions exists if and only if $\dim X \geqq n$.

Definition VI 1. Suppose f is a mapping of a space X in a space Y. A point y of $f(X)$ is called an *unstable value* of f if for every positive δ there is a mapping g of X in Y satisfying

$$(1) \qquad d(f(x), g(x)) < \delta \qquad \text{for every } x \text{ in } X,$$

$$(2) \qquad g(X) \subset Y - y.$$

Other points of $f(X)$ are called *stable* values of f.

Example VI 1. Let f be the mapping of the straight line in itself defined by the function $y = x^2$. The point $y = 0$ is an unstable value of f. All other values $y > 0$ are stable.

Example VI 2. Let f be the identity mapping of the n-cube I_n on itself. Every boundary point of I_n is an unstable value of f. For I_n can be transformed, by a mapping differing very little from the identity, into a smaller concentric cube. On the other hand, every interior point of I_n is a stable value of f. It is sufficient, on grounds of homogeneity, to prove this for the origin $(0, 0, \cdots, 0)$. We shall show that for any mapping g satisfying (1) with $\delta = \frac{1}{2}$ the origin is a point of $g(I_n)$. Using vectorial notation, we consider the mapping

$$k(x) = x - g(x) = f(x) - g(x).$$

This mapping transforms the cube $|x| \leqq \frac{1}{2}$ into itself and hence by Brouwer's Fixed-Point Theorem (page 40) there must be a point x_0 for which $k(x_0) = x_0$, i.e. $g(x_0) = 0$.

Example VI 3. Let X be an arbitrary set in E_n and f the identity mapping of X in E_n. Then every interior point p of X is a stable value of f. For p is contained in a cube Q contained in X and from the previous example it follows that p is a stable value for the partial* mapping $f|Q$, and therefore a fortiori for f. We leave to the reader the proof that the boundary points of X are unstable.

Example VI 4. Suppose X is an arbitrary space and f a mapping of X in I_n. Then every point on the boundary of I_n is unstable. For given any positive $\delta(< 1)$, the functions

$$g_i(x) = (1 - \delta)f_i(x), \qquad i = 1, \cdots, n. \tag{3}$$

define a mapping whose image covers no boundary point of I_n.

Theorem VI 1. *Let X be a space of dimension less than n and f a mapping of X in I_n. Then all values of f are unstable.*

Proof. We know from Example VI 4 that no boundary point of I_n can be a stable value. Hence it is sufficient to prove that there are no stable points in the interior of I_n, and this amounts to the assertion that the origin is not a stable value of f. Let $f(x)$ have the coordinates $f_1(x), \cdots, f_n(x)$. Let δ be an arbitrary positive number, C_i^+ the set of points of X for which

$$f_i(x) \geqq \delta,$$

* The mapping f considered as operating only on Q.

and C_i^- the set of points of X for which

$$f_i(x) \leqq -\delta.$$

For each i the sets C_i^+ and C_i^- are closed and disjoint. Hence by III 5 C) there exist closed sets $B_1, \cdots, B_n$ such that B_i separates C_i^+ and C_i^-, i.e.

$$X - B_i = U_i^+ + U_i^-.$$

U_i^+ and U_i^- being disjoint open sets which contain C_i^+ and C_i^- respectively, and

(4) $$B_1 \cdots B_n = 0.$$

We define new functions $g_1(x), \cdots, g_n(x)$:

$$g_i(x) = f_i(x) \qquad \text{if} \quad x \in C_i^+ + C_i^-$$

$$g_i(x) = \delta \frac{d(x, B_i)}{d(x, C_i^+) + d(x, B_i)} \qquad \text{if} \quad x \in U_i^+ - C_i^+$$

$$g_i(x) = -\delta \frac{d(x, B_i)}{d(x, C_i^-) + d(x, B_i)} \qquad \text{if} \quad x \in U_i^- - C_i^-$$

$$g_i(x) = 0 \qquad \text{if} \quad x \in B_i.$$

One readily sees that the $g_i(x)$ are continuous and that

(5) $$|g_i(x) - f_i(x)| \leqq 2\delta.$$

Moreover, $g_i(x)$ is zero only when x is in B_i. Hence by (4) there is no point in X at which all the functions $g_i(x)$ vanish simultaneously. This means that the origin is not an image point under the mapping defined by the g_i and, with (5), this shows that the origin is not a stable value of f.

A) Given a mapping f of a space X in I_n and a point $y \in I_n$, if

$$f(X) \subset I_n - y$$

there exists for every positive δ a mapping g of X in I_n such that

(6) $$d(f(x), g(x)) < \delta \qquad \text{for every } x \text{ in } X,$$

(7) $$\overline{g(X)} \subset I_n - y.$$

PROOF. If y lies on the boundary of I_n the mapping determined by (3) in Example VI 4 is adequate. If y is an interior point of I_n let $f'(x)$ be the projection of the point $f(x)$ from y on the boundary of I_n.

If the length of the segment joining $f(x)$ and $f'(x)$ is $\geqq \frac{1}{2}\delta$ we define $g(x)$ to be the point on this segment at distance $\frac{1}{2}\delta$ from $f(x)$; if the length of the segment joining $f(x)$ and $f'(x)$ is less than $\frac{1}{2}\delta$ we put $g(x)$ equal to $f'(x)$. Evidently $g(x)$ fulfills (6) and (7). This proves A).

We now prove the converse of Theorem VI 1 by making use of the technique of functional spaces developed in Chapter V.

Theorem VI 2. *If X is a space of dimension $\geqq n$ there exists a mapping of X in I_n with at least one stable value.*

Proof. Suppose to the contrary that no mapping of X in I_n has stable values. It follows, from A) and the definition of unstable values, that for each point y in I_n every mapping f of X in I_n can be approximated arbitrarily closely by mappings g with the property

$$\overline{g(X)} \subset I_n - y.$$

Let us now consider the functional space I_ω^X of the mappings of X in the Hilbert-cube (see Definition V 3). Let $M = M(i_1, \cdots, i_n; c_1, \cdots, c_n)$ be the linear subspace of Hilbert space defined by the n equations

$$x_{i_1} = c_1, \cdots, x_{i_n} = c_n. \tag{8}$$

We denote by $G(M)$ the set of mappings g of X in I_ω with the property

$$\overline{g(X)} \subset I_\omega - M.$$

The set $G(M)$ is *dense* in I_ω^X. For given an arbitrary mapping

$$f(x) = (f_1(x), f_2(x), \cdots), \qquad |f_i(x)| \leqq \frac{1}{i},$$

of X in I_ω the functions

$$f_{i_1}(x), \cdots, f_{i_n}(x) \tag{9}$$

define a mapping of X in I_n, and as remarked above, arbitrarily small changes of the mapping (9) suffice to free the point $(c_1, \cdots, c_n)$ from the closure of the image of X.

Furthermore, the set $G(M)$ is *open* in I_ω^X. For if g is in $G(M)$ the distance

$$d = d(g(X), M)$$

is positive; every mapping f with $d(f, g) < d$ then belongs to $G(M)$.

We now fix our attention on those mappings g of X in I_ω for which

(10) $$\overline{g(X)} \subset \mathfrak{M}_\omega^{n-1},$$

$\mathfrak{M}_\omega^{n-1}$ denoting (see Example III 7) the set of points in I_ω at most $n - 1$ of whose coordinates are rational. The complement of $\mathfrak{M}_\omega^{n-1}$ in I_ω is the intersection of I_ω with the sum of a countable number of hyperplanes $M_1, M_2, \cdots$ of type (8), namely those corresponding to all possible combinations of n indexes i_j and n rational numbers c_j. Hence (10) is equivalent to $\overline{g(X)} \subset I_\omega - M_i$, or

$$g \in G(M_i), \qquad \text{for each } i = 1, 2, \cdots.$$

$I_\omega{}^X$ contains, by Theorem V 4, a dense G_δ-set H each of whose elements is a homeomorphism. The set

$$H' = H \prod_i G(M_i), \qquad i = 1, 2, \cdots,$$

as a countable intersection of dense G_δ-sets in a complete space, is itself dense in $I_\omega{}^X$ (see V 2 B)); in particular it is not empty.

Hence there is a homeomorphism h transforming X in a subset of $\mathfrak{M}_\omega^{n-1}$. But (Example III 7)

$$\dim \mathfrak{M}_\omega^{n-1} \leqq n - 1$$

in contradiction with the assumption that

$$\dim X \geqq n.$$

This completes the proof of Theorem VI 2.

REMARK. Theorem VI 2 implies that in a space of dimension $\geqq n$ it is always possible to define n pairs of closed sets C_i, C_i', $i = 1, \cdots, n$, with $C_iC_i' = 0$ satisfying the following condition: If B_i, $i = 1, \cdots, n$, is a closed set separating C_i and C_i', then $B_1 \cdots B_n \neq 0$. For otherwise we could apply the argument in the proof of Theorem VI 1 to contradict Theorem VI 2.

The following proposition shows that for mappings in I_n the question of whether a point is stable or unstable depends only on the behavior of the mapping in arbitrarily small neighborhoods of the point.

B) Let f be a mapping of a space X in I_n. An interior* point y of

* The adjective "interior" may be omitted since it can very easily be shown that both hypothesis and conclusion are always satisfied, for any mapping, for boundary points.

$f(X)$ is an unstable value of f if and only if for every neighborhood U of y there exists a mapping g of X in I_n satisfying

$$(11) \qquad g(x) = f(x) \quad \text{if} \quad f(x) \notin U,$$

$$(12) \qquad g(x) \in U \quad \text{if} \quad f(x) \in U,$$

$$(13) \qquad y \notin g(X).$$

PROOF. One sees without difficulty that the condition is sufficient. For from (11) and (12) we have

$$d(f(x), g(x)) \leqq \delta(U) \quad \text{for every } x \text{ in } X,$$

so that there are mappings g approximating f arbitrarily closely and satisfying (2).

To prove the necessity let y be an interior point of I_n and δ a positive real number. Without loss of generality we may assume that y is the origin and U is a spherical neighborhood of y of radius δ. Since y is an unstable value of f there exists a mapping g' of X in I_n for which, in vectorial notation,

$$(14) \qquad |f(x) - g'(x)| = d(f(x), g'(x)) < \tfrac{1}{2}\delta,$$

$$(15) \qquad g'(x) \neq 0.$$

We construct a new mapping g as follows:

$$(16) \qquad g(x) = g'(x) \qquad \text{if} \quad |f(x)| \leqq \tfrac{1}{2}\delta,$$

$$(17) \qquad g(x) = 2\left(1 - \frac{|f(x)|}{\delta}\right)g'(x) - \left(1 - \frac{2|f(x)|}{\delta}\right)f(x)$$

$$\text{if} \quad \tfrac{1}{2}\delta < |f(x)| < \delta,$$

$$(18) \qquad g(x) = f(x) \qquad \text{if} \quad |f(x)| \geqq \delta.$$

We verify that $g(x)$ is a mapping of X in I_n with the properties (11)–(13). (11) is the same as (18). For $\frac{1}{2}\delta < |f(x)| < \delta$ we get by a simple computation from (14) and (17):

$$|f(x) - g(x)| = 2\left(1 - \frac{|f(x)|}{\delta}\right)|f(x) - g'(x)| < \delta - |f(x)|$$

and hence

$$(19) \qquad 0 < |g(x)| < \delta.$$

By (14), (15), and (16) the inequality (19) is still valid for $|f(x)| \leqq \frac{1}{2}\delta$. This proves (12) and (13), and completes B).

From B) we get the following extension of Theorem VI 1:

C) Let X be a space of dimension less than n and f a mapping of X in a space B containing an open subset U homeomorphic to E_n. Then all values of f contained in U are unstable.

PROOF. Let U' be the inverse-image under f of U. The partial mapping $f|U'$ can be regarded as a mapping of U' in I_n since U is homeomorphic to the interior of I_n. Let $y \varepsilon U$ and let V be a neighborhood of y such that $\overline{V} \subset U$. By Theorem VI 1 and B) there exists a mapping g of U' in U such that

(20) $$g(x) = f(x) \quad \text{if} \quad f(x) \notin V,$$

(21) $$g(x) \varepsilon V \quad \text{if} \quad f(x) \varepsilon V,$$

(22) $$y \notin g(X).$$

By putting $g(x)$ equal to $f(x)$ for $x \varepsilon X - U'$ we obtain a mapping, which we continue to denote by g, of all of X in B. The new mapping g retains properties (20), (21), (22). Since V, and hence $d(f, g) = \sup_{x \varepsilon X} d(f(x), g(x))$ can be made arbitrarily small, it follows that every value of f is unstable.

D) If f is a mapping of X in an n-sphere and $\dim X < n$ then all values of f are unstable.

PROOF. Proposition D) is a corollary to Proposition C).

2. Extensions of mappings

Definition VI 2. Let A be a subset of a space X and $f(x)$ a mapping of A in a space Y. A mapping $F(x)$ of X in Y satisfying

$$F(x) = f(x) \quad \text{for} \quad x \text{ in } A$$

is called an *extension of f over X with respect to Y*. When no confusion is possible the words "with respect to Y" will be omitted.

Theorem VI 3. Tietze's Extension Theorem. *Suppose C is a closed subset of a space X and f(x) a continuous real-valued function defined over C and bounded by a constant k:*

$$|f(x)| \leqq k,$$

Then there exists a real-valued extension F(x) of f(x) over X such that

$$|F(x)| \leqq k.$$

Proof.* We demonstrate first the existence of a continuous function F_1 defined over X satisfying

$$|F_1(x)| \leqq \tfrac{1}{3}k, \quad x \,\varepsilon\, X,$$
$$|F_1(x) - f(x)| \leqq \tfrac{2}{3}k, \quad x \,\varepsilon\, C.$$

Let C^+ be the set of points of C for which

$$f(x) \geqq \tfrac{1}{3}k.$$

and C^- the set for which

$$f(x) \leqq -\tfrac{1}{3}k.$$

Then the function

$$F_1(x) = \tfrac{1}{3}k \frac{d(x, C^-) - d(x, C^+)}{d(x, C^-) + d(x, C^+)}$$

has the required properties, since

$$\begin{aligned} F_1(x) &= \tfrac{1}{3}k && \text{if } f(x) \geqq \tfrac{1}{3}k \\ -\tfrac{1}{3}k \leqq F_1(x) &\leqq \tfrac{1}{3}k && \text{if } -\tfrac{1}{3}k \leqq f(x) \leqq \tfrac{1}{3}k, \\ F_1(x) &= -\tfrac{1}{3}k && \text{if } f(x) \leqq -\tfrac{1}{3}k. \end{aligned}$$

Replacing $f(x)$ by $f(x) - F_1(x)$ and k by $\tfrac{2}{3}k$ we define over X a function $F_2(x)$ such that

$$|F_2(x)| \leqq \frac{2}{3^2}k, \quad x \,\varepsilon\, X,$$

$$|f(x) - F_1(x) - F_2(x)| \leqq \frac{2^2}{3^2}k, \quad x \,\varepsilon\, C.$$

Continuing in the same way we get a sequence $\{F_n(x)\}$ of continuous functions over X which satisfy

$$(1) \qquad |F_n(x)| \leqq \frac{2^{n-1}}{3^n}k, \quad x \,\varepsilon\, X$$

$$(2) \qquad \left| f(x) - \sum_{i=1}^{n} F_i(x) \right| \leqq \frac{2^n}{3^n}k, \quad x \,\varepsilon\, C.$$

(1) shows that the series

* This proof is due to Urysohn: Über die Mächtigkeit zusammenhängender Mengen, *Math. Ann.* 94 (1925), pp. 262–295, in particular p. 293.

$$\sum_{n=1}^{\infty} F_n(x)$$

converges uniformly over X and therefore has a continuous sum $F(x)$. From (1) and (2) follow the relations

$$|F(x)| \leqq k$$

and

$$F(x) = f(x) \quad \text{for} \quad x \in C,$$

q.e.d.

Corollary 1. *Let C be a closed subset of a space X and f any mapping of C in I_n (in I_ω). Then f can be extended over X (with respect to I_n (to I_ω)).*

Proof. This follows from application of Theorem VI 3 to each of the coordinates of $f(x)$.

If we replace I_n by the n-sphere S_n and consider a mapping f of a closed subset C of X in S_n, then, in general, f cannot be extended over all of X with respect to S_n. For example let X be the closed spherical region $\sum_{i=1}^{n+1} x_i^2 \leqq 1$ in E_{n+1}, which is bounded by S_n. As has been demonstrated in IV 1 B) the identity mapping of S_n on S_n cannot be extended to a mapping of X in S_n. However:

Corollary 2. *Let C be a closed subset of a space X and f a mapping of C in the n-sphere S_n. Then there is an open set in X containing C over which f can be extended (with respect to S_n).**

Proof. Let the coordinates of $f(x)$ in E_{n+1} be

$$f_1(x), \cdots, f_{n+1}(x).$$

If S_n has radius 1,

$$\sum_{i=1}^{n+1} (f_i(x))^2 = 1,$$

and consequently

$$|f_i(x)| \leqq 1, \qquad i = 1, \cdots, n+1.$$

* Corollaries 1 and 2 express certain properties of the spaces I_n and S_n respectively.

If a space A has the property proved in Corollary 1 for I_n it is called an *absolute retract*; if A has the property proved in Corollary 2 for S_n it is called an *absolute neighborhood retract.* It can easily be shown that every polytope is an absolute neighborhood retract. These notions were introduced by Borsuk and have played a prominent role in the topological research of recent years.

Tietze's Extension Theorem yields extensions $F_i(x)$ of $f_i(x)$ over X. Let U be the set of points in X for which

$$\sum (F_i(x))^2 > 0.$$

U is quite clearly an open set containing C. If we let $F(x)$ be the point in S_n whose i^{th} coordinate is

$$\frac{F_i(x)}{\left[\sum_{i=1}^{n+1} (F_i(x))^2\right]^{1/2}}$$

the mapping F will be the desired extension of f over U.

Tietze's Extension Theorem makes it possible to rephrase Theorems VI 1 and VI 2 in terms of mappings in spheres:

Theorem VI 4. *A space X has dimension $\leqq n$ if and only if for each closed set C and mapping f of C in S_n there is an extension of f over X.*

PROOF. The condition is necessary. We are given a closed set C and a mapping f of C in S_n, which we here take as the boundary of I_{n+1}. f is then a mapping of C in I_{n+1}. By Corollary 1 to Tietze's Extension Theorem there exists a mapping F' of X in I_{n+1} which is an extension of f. Because

$$\dim X \leqq n$$

Theorem VI 1 implies that the origin is not a stable value of F'. Then 1 B) furnishes a mapping F'' of X in I_{n+1} such that the origin is not contained in $F''(X)$ while $F''(x) = F'(x)$ for all values $F'(x)$ not in the interior of I_{n+1}. In particular, for x in C,

$$F''(x) = F'(x) = f(x).$$

Let $F(x)$ be the projection of the point $F''(x)$ on the boundary of I_{n+1} from the origin. Then F is plainly the desired extension of f.

The condition is sufficient. In order to prove that $\dim X \leqq n$ it is enough to show, according to Theorem VI 2, that a mapping f of X in I_{n+1} cannot have stable values. A boundary point of I_{n+1} is never stable (see Example VI 4). Hence let y be an interior point of I_{n+1}, U a spherical neighborhood of y of radius $\frac{1}{2}\delta$, and let us take S_n as the boundary of U. Denote by C the inverse-image under f of S_n. C is closed because f is continuous. By hypothesis there is a mapping F of the entire space X in S_n such that $F(x) = f(x)$ for x in C. Now construct a new mapping $g(x)$ of X in I_{n+1} according to the rules

$$g(x) = f(x) \quad \text{if} \quad f(x) \notin U,$$
$$g(x) = F(x) \quad \text{if} \quad f(x) \in U.$$

$g(x)$ is a mapping of X in $I_{n+1} - U$ and $d(f, g) < \delta$. This proves the theorem.

Corollary. *Let C be a closed subset of a space X. Then if $X - C$ has dimension $\leqq n$ every mapping of C in S_n can be extended over X.*

Proof. Let f be a mapping of C in S_n. Corollary 2 of Tietze's Extension Theorem shows that there is an open set U containing C and an extension f' of f over U. Let V be an open set in X satisfying

$$C \subset V \subset \overline{V} \subset U;$$

V exists because of the normality* of X. Consider the partial mapping

$$f' \mid \overline{V} \cdot (X - C).$$

This is a mapping in S_n of a closed subset of the space $X - C$. Now $X - C$ has dimension $\leqq n$. By Theorem VI 4 there is an extension f'' of f' over $X - C$. If we put

$$F(x) = f(x) \quad \text{for} \quad x \in C,$$
$$F(x) = f''(x) \quad \text{for} \quad x \in X - C,$$

we obtain the sought-for extension of f over X.

An important application of Theorem VI 4 will be a simple proof of Brouwer's Theorem on the Invariance of Domain. The reader may, if he wishes, immediately proceed to this theorem, Theorem VI 9.

3. Homotopy

Definition VI 3. Let X and Y be two spaces. We say that a mapping f of X in Y is *homotopic* to a mapping g of X in Y if one can find a function $f(x, t)$ of two variables x and t, x being a point of X and t a real number $0 \leqq t \leqq 1$, which has its values in Y, is continuous in the pair (x, t), and satisfies

$$f(x, 0) = f(x)$$
$$f(x, 1) = g(x).$$

A function $f(x, t)$ of the sort specified above is of course the same thing as a mapping in Y defined over $X \times I$ (the product of X and the unit segment). The intuitive meaning of homotopy is that g can

* See index.

be gotten from f by a process of continuous deformation, the various phases of the deformation all having their images in Y.

It is manifest that the relation of homotopy is both symmetric and transitive, i.e. if f is homotopic to g then g is homotopic to f, and if f is homotopic to g and g to h, then f is homotopic to h. Consequently homotopy effects a division of all the mappings of X in Y into disjoint *homotopy classes.*

Definition VI 4. A mapping of a space X in a space Y is said to be *inessential* if it is homotopic to a constant mapping. A mapping which is not inessential is called *essential.*

Example VI 5. Suppose both X and Y are the n-sphere S_n and f is the identity mapping. Then as Proposition IV 1 A) has proved, f is essential, and this fact served as the basis for the proof that E_n is n-dimensional. It is known that there are countably many homotopy classes of mappings of S_n in itself (Example VIII 25). Each of these is characterized by an integer called its degree; roughly speaking the degree is the algebraic number of times the sphere is wrapped around itself by the mapping.

Example VI 6. If Y is arcwise connected* then for any space X all inessential mappings of X in Y are in the same homotopy class, for arcwise connectedness clearly implies that all constant mappings are homotopic.

Example VI 7. We shall say of a space X that it is *contractible* if the identity mapping, of X on X, is inessential. Intuitively this means that X can be shrunk in itself to a single point. E_n and I_n are examples of contractible spaces; S_n is non-contractible (see Example VI 5). The reader will easily prove that if either X or Y is contractible then all mappings of X in Y are inessential. In particular a mapping of I_n in an arbitrary space is inessential.

Example VI 8. Suppose f and g are two mappings of an arbitrary space X in S_n such that for any $x \varepsilon X$ the points $f(x)$ and $g(x)$ have distance less than the diameter of S_n, i.e. are never antipodal. Then f and g are homotopic; for there is always defined uniquely the minor arc of the great circle joining $f(x)$ and $g(x)$ and we may take $f(x, t)$ as the point dividing this arc in the ratio $t/1 - t$. As a consequence we have: Any mapping f of X in S_n leaving a point q of S_n free from $f(X)$ is inessential. For by what we have just stated, f is homotopic to the constant mapping g of X in the antipode of q.

* See index.

From now on we shall consider mainly mappings in S_n. We return to the question of extending a mapping defined over a closed subset of a space to a mapping defined over the whole space. It turns out that the existence of such an extension depends only on the homotopy class of the given mapping, a fact having many very important applications.

Theorem VI 5. Borsuk's Theorem.* *Let C be a closed subset of a space X and f and g two homotopic mappings of C in S_n. Then if there is an extension F of f over X there is also an extension G of g over X, with F and G homotopic.*

We interpose a Proposition concerning open sets in a product space.

A) Let C be a subset of a space X. In the product† space $X \times I$ let U be an open set containing $C \times I$. Then there is an open set V in X containing C such that $V \times I$ is contained in U.

PROOF. First we show that each point c of C has a neighborhood v in X for which $v \times I \subset U$. Consider the segment $c \times I$. This is contained in U. Each of its points therefore, has a neighborhood contained in U of the form $w \times i$, where w is a neighborhood in X of c and i is an interval in I. Because $c \times I$ is compact a finite number of these "rectangular" neighborhoods cover $c \times I$. We take v as the intersection of the projections on X of these neighborhoods. Proposition A) itself is established by taking V as the sum, over all the points of C, of the neighborhoods v.

We now return to the

PROOF OF BORSUK'S THEOREM. The homotopy of f and g means that there is a mapping $f(x, t)$ of $C \times I$ in S_n satisfying

$$f(x, 0) = f(x)$$
$$f(x, 1) = g(x),$$

x being restricted to C. There also exists, by hypothesis, a mapping F of X in S_n coinciding with f on C. Let C' be the set in $X \times I$ consisting of the points $(x, 0)$ for $x \in X$, and the points (x, t) for $x \in C$ and $0 \leqq t \leqq 1$. C' is a closed subset of $X \times I$. We consider the following mapping of C' in S_n:

* Note that the only property of S_n used in the proof is that of being an absolute neighborhood retract (see footnote on page 82). Hence Borsuk's Theorem remains true if S_n is replaced by an arbitrary absolute neighborhood retract. The proof given here is due to C. H. Dowker: Mapping Theorems for Noncompact Spaces, *Am. J. Math.* 69 (1947), pp. 200–242, especially p. 232.

† I is the unit segment [0, 1].

$$F(x, 0) = F(x) \quad \text{for} \quad x \,\varepsilon\, X$$

and

$$F(x, t) = f(x, t) \quad \text{for} \quad x \,\varepsilon\, C, \qquad 0 \leqq t \leqq 1.$$

By Cor. 2, Thm. VI 3 there is an open set $U \supset C'$ in $X \times I$ over which $F(x, t)$ can be extended; we continue to denote the extended mapping by $F(x, t)$. By A) there is an open set V in X containing C for which $V \times I$ is contained in U. Note that $F(x, t)$ is defined for any $x \,\varepsilon\, V$ and $0 \leqq t \leqq 1$, and furthermore for any $x \,\varepsilon\, X$ and $t = 0$.

C and the complement of V are disjoint closed subsets of X. Hence there is a continuous real valued function $p(x)$ defined† over X whose range is between 0 and 1 and which is 1 on C and 0 on the complement of V. Now consider the function

$$G(x, t) = F(x, tp(x)).$$

$G(x, t)$ is defined for all $x \,\varepsilon\, X$ and $0 \leqq t \leqq 1$, and is continuous in (x, t). If we define $G(x)$ by

$$G(x) = G(x, 1)$$

it is clear that

$$G(x) = g(x) \quad \text{for} \quad x \,\varepsilon\, C,$$

so that $G(x)$ is an extension of $g(x)$ over X. It is likewise clear that $G(x, 0) = F(x)$; since $G(x, 1) = G(x)$ by definition, F and G are homotopic.

Corollary. *An inessential mapping of a closed subset of a space X in S_n can always be extended over X.*

Proof. For a constant mapping can always be extended.

We now establish a connection between dimension and homotopy.

B) Suppose f and g are two mappings of a space X in S_n such that the points for which $f(x)$ is not equal to $g(x)$ form a subset D of dimension $\leqq n - 1$. Then f and g are homotopic.

Proof. D is obviously open. Let D^* be the closed set in $X \times I$ consisting of the points $(x, 0)$ and $(x, 1)$ for $x \,\varepsilon\, X$, and the points (x, t) for $x \,\varepsilon\, X - D$ and $0 \leqq t \leqq 1$. We define a mapping $F(x, t)$ of D^* in S_n as follows:

$$F(x, t) = f(x) = g(x) \quad \text{for} \quad x \,\varepsilon\, X - D,$$
$$F(x, 0) = f(x), \qquad F(x, 1) = g(x).$$

† E.g. $p(x) = d(x, X - V)/[d(x, X - V) + d(x, C)]$. See AH, pp. 74, 76.

The complement of D^* is contained in $D \times I$, which has dimension $\leqq n$ by Theorem III 4. By the Corollary to Theorem VI 4 one can extend $F(x, t)$ to a mapping of $X \times I$ in S_n, proving the homotopy of f and g.

Theorem VI 6. *If X is a space of dimension less than n then all mappings of X in S_n are homotopic, and hence inessential.*

PROOF. This is an immediate consequence of B).

By combining Proposition B) with Borsuk's Theorem we obtain the following result, which asserts that if mappings defined on two closed parts of a space fit together except possibly on a set of low dimension, then either of the mappings can be modified so as to fit the other completely.

C) Suppose a space X is the sum of two closed subsets C_1 and C_2. Suppose F_1 and F_2 are mappings of C_1 and C_2 in S_n. Suppose furthermore that the points (in C_1C_2) for which $F_1(x)$ is not equal to $F_2(x)$ form a set of dimension $\leqq n - 1$. Then F_1 can be extended over X.

PROOF. The partial mappings $F_1|C_1C_2$ and $F_2|C_1C_2$ differ only on a set of dimension $\leqq n - 1$ and hence are homotopic by B). As $F_2|C_1C_2$ admits an extension over C_2, namely F_2, it follows from Borsuk's Theorem that there is an extension F_1', say, of $F_1|C_1C_2$ over C_2. Putting

$$F(x) = F_1(x) \quad \text{for} \quad x \in C_1,$$

$$F(x) = F_1'(x) \quad \text{for} \quad x \in C_2,$$

we achieve the desired extension.

We now establish several results which will have important applications in the following sections.

D) Let f and g be mappings of a space X in S_n. Suppose X is the sum of two closed subsets C_1 and C_2 whose intersection has dimension $\leqq n - 2$. If f and g are homotopic on each of C_1 and C_2, i.e. if $f|C_1$ is homotopic to $g|C_1$ and $f|C_2$ is homotopic to $g|C_2$, then f and g are homotopic.

PROOF. Consider the product space $X \times I$. By hypothesis there exist mappings

$$f_1(x, t) \quad \text{defined for} \quad x \in C_1, \qquad 0 \leqq t \leqq 1$$

and

$$f_2(x, t) \quad \text{defined for} \quad x \in C_2, \qquad 0 \leqq t \leqq 1$$

satisfying

$$\text{for } x \in C_1: \quad f_1(x, 0) = f(x), \quad f_1(x, 1) = g(x),$$
$$\text{for } x \in C_2: \quad f_2(x, 0) = f(x), \quad f_2(x, 1) = g(x).$$

We extend the domain of definition of $f_1(x, t)$ by putting

$$\text{for } x \in C_2: \quad f_1(x, 0) = f(x), \quad f_1(x, 1) = g(x).$$

Each of the functions $f_1(x, t)$ and $f_2(x, t)$ is defined over a closed subset of $X \times I$ and the points for which $f_1(x, t)$ and $f_2(x, t)$ are defined but have different values are contained in $(C_1C_2) \times I$, and hence (Theorem III 4) form a set of dimension $\leqq n - 1$. Proposition C) now shows that there is an extension $F(x, t)$ of $f_1(x, t)$ over the whole of $X \times I$. We have

$$F(x, 0) = f_1(x, 0) = f(x)$$
$$F(x, 1) = f_1(x, 1) = g(x),$$

proving that f and g are homotopic.

E) Let f be a mapping of a space X in S_n. Suppose X is the sum of two closed subsets C_1 and C_2 whose intersection has dimension $\leqq n - 2$. If f is inessential on each of C_1 and C_2, i.e. if $f|C_1$ and $f|C_2$ are inessential, then f is inessential.

Proof. This is the special case of D) in which g is a constant mapping.

F) Let C be a closed subset of a space X and $\{V_\lambda\}$ a collection of open sets whose sum is X and whose boundaries have dimension $\leqq n - 1$. If f is a mapping of C in S_n which can be extended over each of the sets $C + \overline{V}_\lambda$ then f can be extended over the whole space X.

Proof. Because X is separable we may assume (AH, p. 78) that $\{V_\lambda\}$ is a countable collection, $V_1, V_2, \cdots$. We define successive extensions of f, letting F_1 be an arbitrary extension of f over $C + \overline{V}_1$. Let us assume that we have already defined F_k as an extension of F_{k-1} over $C + \overline{V}_1 + \cdots + \overline{V}_k$. We divide $C + \overline{V}_1 + \cdots + \overline{V}_{k+1}$ into the two closed parts

$$C_1 = C + \overline{V}_1 + \cdots + \overline{V}_k$$

and

$$C_2 = (C + \overline{V}_{k+1}) - (V_1 + \cdots + V_k).$$

By the hypothesis, f admits an extension F^k over C_2. The points x for which $F_k(x) \neq F^k(x)$ are contained in the set

$$C_1C_2 - C \subset \text{bdry}\,(V_1 + \cdots + V_k) \subset \text{bdry}\,V_1 + \cdots + \text{bdry}\,V_k,$$

which, because of the Sum Theorem (Theorem III 2), has dimension $\leqq n - 1$. Hence we may apply C) to get an extension F_{k+1} of F_k over $C + \bar{V}_1 + \cdots + \bar{V}_{k+1}$.

Now let x be an arbitrary point of X and m the first integer for which $V_m \supset x$. Setting $F(x) = F_m(x) = F_{m+1}(x) = F_{m+2}(x) = \cdots$ we obtain a mapping F of X in S_n which is clearly an extension of f.

G) Suppose a space X is the sum (not necessarily countable)* of a family of closed sets $\{K_\lambda\}$ with these two properties: each K_λ has dimension $\leqq n$, and given any K_λ and open set U containing K_λ there is an open set V,

$$K_\lambda \subset V \subset U,$$

with

$$\dim \text{bdry}\, V \leqq n - 1.$$

Then X has dimension $\leqq n$.

PROOF. By virtue of Theorem VI 4 in order to demonstrate that X has dimension $\leqq n$ it suffices to show that given any closed set C in X and any mapping f of C in S_n there is an extension of f over X. From the fact that $\dim K_\lambda \leqq n$ we know that for each K_λ one can extend f over $C + K_\lambda$ (let $X = C + K_\lambda$ in the Corollary to Theorem VI 4, so that $X - C \subset K_\lambda$), and (Corollary 2 to Theorem VI 3) one can even extend f over an open set U_λ,

$$U_\lambda \supset C + K_\lambda. \tag{1}$$

It follows from the hypotheses that there is an open set V_λ for which

$$K_\lambda \subset V_\lambda \subset \bar{V}_\lambda \subset U_\lambda, \quad \text{and} \tag{2}$$

$$\dim \text{bdry}\, V_\lambda \leqq n - 1. \tag{3}$$

Since by (1) and (2) $C + \bar{V}_\lambda \subset U_\lambda$, it is clear that f can be extended over $C + \bar{V}_\lambda$.

From (2) we have that the sum of the V_λ fills out X; Proposition F) shows that f can be extended over X and completes the proof.

* Compare this proposition with the Sum Theorem (Theorem III 2). On one hand the proposition above is stronger in that no assumption of countability is made, while on the other hand it is weaker in that we must require the closed sets of the family to have the property that they can be surrounded arbitrarily closely by open sets whose boundaries have dimension $\leqq n - 1$. Observe also that if $\{K_\lambda\}$ is the set of all points of X then G) reduces to the ordinary definition of dimension.

4. Mappings which lower dimension

A single-valued transformation f of a space X in a space Y may be defined to be continuous by the requirement that its inverse f^{-1} sends closed sets in Y into closed sets in X, or equivalently (because f is defined throughout X), open sets in Y into open sets in X. This suggests that we say that f^{-1} (a many-valued transformation) is continuous if f itself sends closed sets of X into closed sets of Y. For this reason we might decide to call a continuous transformation f *bicontinuous* if f sends closed sets of X into closed sets of Y. There is an important objection to this, however, in that it is just as reasonable to say that f^{-1} is continuous if f sends *open* sets of X into *open* sets of Y; unfortunately the two possible ways of defining the continuity of f^{-1} do *not* coincide (see Examples VI 9 and 9.1) except for *one to one* transformations of X *on* Y. It is probably best, therefore, to avoid the ambiguous word bicontinuous and speak only of closed and open mappings, as defined below.

Definition VI 5. A *closed* (*open*) mapping of a space X in a space Y is one which sends closed (open) subsets of X into closed (open) subsets of Y.

EXAMPLE VI 9. Let X be the space of real numbers $0 < x < 1$ and Y the space $0 \leqq x \leqq 1$. Let f be the identity mapping $f(x) = x$ of X in Y. Then f is not closed, since X, which is of course closed in X, is not closed in Y. It is easy to see, however, that f is open.

EXAMPLE VI 9.1. Let X be the subset of the plane consisting of the vertical and horizontal axes, and Y the horizontal axis. Let f be the vertical projection of X on Y. Then f is closed, but not open.

REMARK. A closed subset of a compact space is compact (AH, p. 86) and a continuous image of a compact space is compact (AH, p. 95). Furthermore a compact space is closed in every space containing it (AH, p. 91); hence *every* mapping of a compact space is closed.

Consider the orthogonal projection of E_{n+k} into E_n; this mapping lowers dimension by k units. Each point of E_n has an inverse-image of dimension k. We shall now prove, generally, that a closed mapping of one space in another cannot lower dimension by k units unless at least one k-dimensional set is collapsed into a single point.

Theorem VI 7. *Suppose f is a closed* mapping of a space X in a space Y and*

* For compact spaces the word "closed" is redundant: see the remark above. The theorem is not true for open mappings: see Example VI 10 below.

$$\dim X - \dim Y = k,$$

$k > 0$. *Then there is a point of* Y *whose inverse-image has dimension* $\geqq k$.

Proof. For the purposes of the demonstration we prefer to state the theorem in this form: If for every $y \in Y$

$$\dim f^{-1}(y) \leqq m$$

then

$$\dim X \leqq m + \dim Y.$$

Obviously we may assume that Y is finite-dimensional. We shall prove the theorem by induction on dim Y, keeping m fixed. The assertion is trivial if $\dim Y = -1$, for in that case X also is vacuous. Assuming the assertion now for $\dim Y \leqq n - 1$ we shall prove it for $\dim Y = n$.

In order to prove $\dim X \leqq m + n$ it suffices to show that the family of closed sets $f^{-1}(y)$ satisfy the conditions of Proposition 3 G), n being replaced by $m + n$. By hypothesis we have that

$$\dim f^{-1}(y) \leqq m \leqq m + n.$$

Moreover if U is an open set in X containing $f^{-1}(y)$ the set

$$C = f(X - U)$$

is a closed set in Y as the image under the closed mapping f of the closed set $X - U$. This set C does not contain y. By the hypothesis $\dim Y \leqq n$ there exists a neighborhood V of y in Y with

(1) $$C \cdot V = 0$$

(2) $$\dim \text{bdry } V \leqq n - 1.$$

From (1) it follows that

$$f^{-1}(C) \cdot f^{-1}(V) = 0,$$

and consequently

$$f^{-1}(V) \subset U.$$

Now $f^{-1}(V)$ is an open set, as the inverse image under a continuous transformation of an open set, and contains $f^{-1}(y)$. Let B be the boundary of $f^{-1}(V)$. It is easily seen that $f(B)$ is contained in the boundary of V. Hence by (2), $\dim f(B) \leqq n - 1$. We now apply the hypothesis of the induction to conclude that B has dimension $\leqq m + n - 1$; this shows that the conditions of 3 G) are fulfilled.

Example VI 10. Let us go back to Knaster and Kuratowski's totally disconnected space $X - a$ (see Example II 16) which becomes connected upon adjunction of the single point a. Using the notation of Example II 16 consider the mapping f of $X - a$ on $\mathcal{C}$ which sends each $L^*(p)$ into the point p and each $L^*(q)$ into the point q. Now

$$\dim (X - a) = 1$$

(see Example IV 3) and

$$\dim f(X - a) = \dim \mathcal{C} = 0$$

(see Example II 3), so that f lowers dimension by one unit. Nevertheless the inverse-image of each point of $\mathcal{C}$ has dimension 0. This is no contradiction to Theorem VI 7 because f is not closed; as has been noted by J. H. Roberts, f is open.

Remark. There exists a sort of dual to Theorem VI 7 for mappings which raise dimension: Suppose f is a closed mapping of a space X on a space Y and suppose

$$\dim Y - \dim X = k,$$

$k > 0$. Then there is at least one point of Y whose inverse-image contains at least $k + 1$ points. The proof of this theorem requires methods quite different from those used here,† and is therefore omitted.

5. Cantor-manifolds

Definition VI 6. A compact n dimensional space, $n \geqq 1$, is called an *n-dimensional Cantor-manifold*‡ if it cannot be disconnected§ by a subset of dimension $\leqq n - 2$.

Example VI 11. It follows from the Corollaries to Theorem IV 4 that I_n is an n-dimensional Cantor-manifold, and so is S_n, and, more generally, any compact n-dimensional manifold.

A) It is easy to see that a Cantor-manifold is connected, and that

† See W. Hurewicz: Über dimensionserhöhende stetige Abbildungen, *Jour. f. Math.* 169 (1933), pp. 71–78.

‡ The use in this connection of the word "manifold" is unfortunately firmly rooted in the literature. A Cantor-manifold need not be a manifold at all, in the usual sense that a manifold is a connected space with a basis made up of homeomorphs of E_n.

§ See Definition IV 1.

an n-dimensional Cantor-manifold has dimension n at each of its points.

B) From IV 5 A) it follows that a compact n-dimensional space C is an n-dimensional Cantor manifold if and only if it is impossible to have

$$C = C_1 + C_2$$

with C_1 and C_2 closed proper subsets of C and

$$\dim C_1C_2 \leqq n - 2.$$

The fundamental theorem on Cantor-manifolds is the following:

Theorem VI 8. *Any compact n-dimensional space X contains a subset which is an n-dimensional Cantor-manifold.*

First we establish

C) Given a compact space X, a closed subset C, and a mapping f of C in S_n which cannot be extended over X, there exists a closed set K in X such that

(1) f cannot be extended over $C + K$, but

(2) if K' is any proper closed subset of K, then f can be extended over $C + K'$.

PROOF. For consider the family $\{K_\lambda\}$ of closed sets such that

(3) f cannot be extended over $C + K_\lambda$.

$\{K_\lambda\}$ is not empty, since it contains X. *If K^0 is the intersection of a monotonic decreasing sequence $\{K_i\}$, $i = 1, 2, \cdots$, of closed sets in $\{K_\lambda\}$ then K^0 is also in $\{K_\lambda\}$.* For suppose K^0 were not in $\{K_\lambda\}$, i.e. suppose it were possible to extend f over $C + K^0$. Then there would be, by Corollary 2 to Theorem VI 3, an open set U containing $C + K^0$ over which f could be extended. Now at least one, say K_{i_0}, of the K_i is contained in U; otherwise all the K_i would meet the complement T of U; this is, however, impossible, since the compactness of T would then require K^0 itself to meet T. But $K_{i_0} \subset U$ implies that f can be extended over $C + K_{i_0}$, and this contradicts (3).

We see, then, that the family $\{K_\lambda\}$ satisfies the hypotheses of Brouwer's Reduction Theorem.* Application of this theorem yields a closed set K irreducible with respect to (3), i.e. a closed set satisfying (1) and (2).

PROOF OF THEOREM VI 8. Since $\dim X = n$ there is, by Theorem

* See index.

VI 4, a closed subset C of X and a mapping f of C in S_{n-1} which cannot be extended over X. By C) there is a closed set K in X satisfying (1) and (2). We assert that K is an n-dimensional Cantor-manifold. For otherwise (see B)),

$$K = C_1 + C_2,$$

C_1 and C_2 closed proper subsets of K and

$$\dim C_1C_2 \leqq n - 2.$$

From (2) it follows that f can be extended to mappings F_1 and F_2 defined over $C + C_1$ and $C + C_2$ respectively. By 3 C) one can extend F_1 over $C + K$, and hence one can extend f over $C + K$, contradicting (1).

Corollary. *Let X be a compact n-dimensional space and let A be the subset of X made up of the points at which X has dimension n. Then* $\dim A = n$.

Proof. Let C be an n-dimensional Cantor-manifold contained in X. This exists by Theorem VI 8. We know from Proposition A) that C is contained in A. Hence $\dim A = n$.*

6. Invariance of domain in E_n

Let X be a subset of an arbitrary space A. Suppose h is a homeomorphism acting on all of A with values in A. If x is an interior† point of X then $h(x)$ is an interior point of $h(X)$, and conversely. Suppose, however, the homeomorphism h is defined only on X. It is by no means true that even in this case h carries interior points into interior points and boundary points into boundary points:

Example VI 12. Let A be the subset of E_3 consisting of the (x_1, x_2)-plane and the x_3-axis, X the subset of A consisting of the x_3-axis, and h a homeomorphism mapping the x_3-axis on the x_1-axis. Then the point (0, 0, 1) is an interior point of X but its image is not an interior point of $h(X)$, since there is no set open in A containing the image-point and contained in $h(X)$.

However if A is a Euclidean space we have

Theorem VI 9. Brouwer's Theorem on the Invariance of Domain.

* If X is *not compact* $\dim A$ may be less than n. One can prove, however, that $\dim A$ is always $\geqq n - 1$ (see Menger: *Dimensionstheorie*, B. G. Teubner, Leipzig, 1928, p. 135.

† See index.

Let X be an arbitrary subset of E_n and h a homeomorphism of X on another subset $h(X)$ of E_n. Then if x is an interior point of X, $h(x)$ is an interior point of $h(X)$. In particular, if A and B are homeomorphic subsets of E_n and A is open, then B is open.

We can assume that X is compact, for an interior point of an arbitrary set X is also an interior point of a compact set CX, e.g. the closure of a sufficiently small neighborhood of the point.

PROOF. We shall prove Theorem VI 9 by characterizing the interior points of X, or what amounts to the same, the boundary points of X, by *intrinsic* topological properties of X, i.e. topological properties involving only the points of X itself:

A) Let X be a compact subset of E_n and x a point of X. Then x is a boundary point of X, $x \varepsilon \overline{(E_n - X)}X$, if and only if x has arbitrarily small neighborhoods U in† X with the property that any mapping of $X - U$ in S_{n-1} can be extended over X (with respect to S_{n-1}).

PROOF. The condition is necessary. Let x be a boundary point of X, $S(x)$ a spherical neighborhood of x in E_n and $U = X \cdot S(x)$. We shall show that U has the desired property. Let B denote the $(n - 1)$-sphere which is the boundary of $S(x)$. By the Corollary to Theorem VI 4 any mapping f of $X - U$ in S_{n-1} can be extended* to a mapping f' defined over $X - U + B$. Let q be a point of $S(x)$ not in X. For each x of X denote by x' the projection of x on B from q. Now define

$$F(x) = f'(x') \quad \text{for} \quad x \varepsilon U,$$

$$F(x) = f(x) \quad \text{for} \quad x \varepsilon X - U.$$

$F(x)$ is the desired extension.

The condition is sufficient. For suppose x were an interior point of X. Let $S(x)$ be a spherical neighborhood of x whose closure is contained in X. We shall show that for any neighborhood U of x contained in $S(x)$ there is a mapping of $X - U$ in S_{n-1} which cannot be extended over X. We identify S_{n-1} with the boundary of $S(x)$ and take for f the projection of $X - U$ on S_{n-1} from x. Then f cannot be extended over X, for this would yield a mapping of the closure of $S(x)$ on its boundary leaving the points of its boundary fixed, and thus contradicting IV 1 B). This completes the proof of Proposition A) and hence of Brouwer's Invariance Theorem.

COROLLARY. *Theorem* VI 9 *remains true if E_n is replaced by an arbitrary manifold.*

† That is, subsets of X which are open in X and contain x.

* Because as a compact set, $X - U$ is closed in every containing space; AH, p. 91.

Proof. For every point of X has a neighborhood in the n-dimensional manifold which is homeomorphic to E_n.

Remark. Theorem VI 9 includes the classical theorem on "Invariance of Dimension of Euclidean Spaces" (already proved by III 1A) and Theorem IV 1): E_n and E_m are not homeomorphic if $n \neq m$. For let $n > m$ and regard E_m as a subspace of E_n. The subset E_m is not open in E_n while E_n, of course, is. Hence, by the Invariance of Domain, there can exist no homeomorphism of E_m on E_n.

7. Separating sets in E_n

From now on we assume that $n \geqq 2$ and proceed to discuss separation theorems in E_n.

We shall use the following notation. Let S_{n-1} be the $(n-1)$-sphere in E_n of radius 1 and center at the origin 0. For each point p in E_n we denote by π_p the mapping of $E_n - p$ in S_{n-1} defined by the rule: $\pi_p(x)$, $x \varepsilon E_n - p$, is the projection of the point $x - p$ (in vector terminology) from 0 on S_{n-1}.

A) Let U be a bounded open set in E_n, p a point in U, and C the boundary of U. Then the partial mapping $\pi_p | C$ cannot be extended over $U + C = \overline{U}$.

Proof. There is no loss of generality in taking p to be the origin. Let r be so large that $U + C$ is contained in the spherical region $S(0, r)$ of radius r with 0 as center. Suppose it were possible to extend $\pi_0 | C$ over $U + C$, to a function φ, say. The formulas

$$\psi(x) = \varphi(x) \quad \text{for} \quad rx \varepsilon U,$$
$$\psi(x) = \pi_0(x) \quad \text{for} \quad rx \varepsilon S(0, r) - U,$$

would then define ψ as a mapping of $S(0, 1)$ on its boundary S_{n-1}. Moreover, for each $x \varepsilon S_{n-1}$ we should have $\psi(x) = \pi_0(x) = x$, contradicting IV 1 B).

Theorem VI 10. *Let C be a compact (= closed bounded) set in E_n. Two points p, q, neither contained in C, are separated by C if and only if the mappings $\pi_p | C$ and $\pi_q | C$ belong to different homotopy classes.*

Proof. Assuming first that p and q are separated by C, we shall prove that $\pi_p | C$ and $\pi_q | C$ are not homotopic. We are given that

$$E_n - C = U + V,$$

U, V being disjoint sets which are open in $E_n - C$ and therefore in E_n, and $p \varepsilon U$, $q \varepsilon V$.

One of the sets U, V is bounded.* For let I_n be an n-cube large enough to contain C. Then, since $E_n - I_n$ is a connected subset of $E_n - C$ it must be contained in either U or V. As a matter of notation we assume that $E_n - I_n \subset V$. It follows that $U \subset I_n$, i.e. U is bounded.

Now $\pi_q | C$ can be extended over $U + C$, in fact over $E_n - q \supset U + C$. On the other hand, according to Proposition A) it is not possible, because the boundary of U is contained in C, to extend $\pi_p | C$ over $U + C$. Hence $\pi_p | C$ and $\pi_q | C$ are not homotopic, since Borsuk's Theorem (p. 86) would be contradicted if they were.

Now let us assume that p and q are not separated by C. Then by a well known property of Euclidean spaces one can join p and q by a continuous arc in $E_n - C$, i.e. one can find a continuous function $f(t)$ of the real parameter t, $0 \leqq t \leqq 1$, with values in $E_n - C$ such that

$$f(0) = p, \qquad f(1) = q.$$

The function

$$\pi_{f(t)}(x), \qquad x \varepsilon C,$$

is then a continuous function in (x, t) demonstrating the homotopy of $\pi_p | C$ and $\pi_q | C$.

COROLLARY. *If the points p and q in E_n are separated by a set $C_1 + C_2$, where C_1 and C_2 are compact sets whose intersection has dimension $\leqq n - 3$, then either C_1 or C_2 separates p and q.*

PROOF. For otherwise $\pi_p | C_1$ would be homotopic to $\pi_q | C_1$ and $\pi_p | C_2$ to $\pi_q | C_2$. Hence by 3 D) the mappings $\pi_p | C_1 + C_2$ and $\pi_q | C_1 + C_2$ would be homotopic, so that, by Theorem VI 10, $C_1 + C_2$ would not separate p and q.

Theorem VI 11. *Suppose C is a compact subset of E_n which separates p and q while no proper closed subset of C does so (an "irreducible" separating set). Then C is an $(n - 1)$-dimensional Cantor-manifold.*

PROOF. Observe first that C contains no non-empty open subset. For otherwise the boundary of C would be a proper closed subset of C separating p and q.

Hence, by Theorem IV 3, dim $C \leqq n - 1$. Furthermore by the Corollary to Theorem VI 10, if

$$C = C_1 + C_2,$$

* The reader will observe that this statement is false for $n = 1$. Nevertheless Theorem VI 10 is trivially true for $n = 1$, as is also Theorem VI 12.

with each of C_1, C_2 a closed proper subset of C then we must have

$$\dim C_1C_2 \geqq n - 2.$$

This shows, 5 B), that C is an $(n - 1)$-dimensional Cantor-manifold.

Remark. Theorem VI 11 can also be formulated as follows: *If a compact subset C of E_n is the common boundary of two disjoint open connected subsets A_1 and A_2 then C is an $(n - 1)$-dimensional Cantor-manifold.* For let p_1 and p_2 be points in A_1 and A_2 respectively. Obviously p_1 and p_2 are separated by C. However they are not separated by any proper closed subset C' of C. For let q be a point in $C - C'$, and denote by U a spherical neighborhood of q such that $U \subset E_n - C'$. Then U contains points of both A_1 and A_2, and hence $A_1 + U + A_2$ is a connected subset of $E_n - C'$ containing p_1 and p_2; thus p_1 and p_2 are not separated by C'.

Theorem VI 12. *Let X be a compact (= closed bounded) subset of E_n and C a closed subset of X. In order that there exist a mapping f of C in S_{n-1} which cannot be extended over X it is necessary and sufficient that there exist a non-empty open subset of E_n which is contained in $X - C$ while its boundary is contained in C.*

Remark. We interpose a remark to readers familiar with homology theory. There is a very close connection between extension properties of mappings on one hand, and homology properties of sets on the other. This connection is expressed as follows (Cor. 3 to Theorem VIII 1′): *Given a compact space X of dimension $\leqq n$ and a closed subset C, every mapping f of C in S_{n-1} can be extended over X if and only if every $(n - 1)$-cycle* mod 1 *in C bounds in C whenever it bounds in X.* If X is the Euclidean n-cube then every $(n - 1)$-cycle in C bounds in X; in this case the previous theorem consequently reduces to a simpler form: In order to be able to extend every mapping f of C in S_{n-1} over I_n it is necessary and sufficient that every $(n - 1)$-cycle in C bound in C. Comparing this with Theorem VI 12 we conclude that in I_n there is complete equivalence between bounding in the point-set sense and bounding in the combinatorial sense of homology theory, i.e. if a closed subset of I_n is a boundary in the point-set sense then it carries an essential $(n - 1)$-cycle which bounds in the combinatorial sense, and conversely. This is exactly the deeper meaning of Theorem VI 12.

Proof of Theorem VI 12. *Necessity.* Suppose f is a mapping of C in S_{n-1} which cannot be extended over X. Then by 5 C) there is a closed set K of X with the properties

(1) f cannot be extended over $C + K$, but

(2) if K' is any proper closed subset of K then f can be extended over $C + K'$.

We now assert that

a) $$K - C \neq 0,$$

b) $$K - C \subset X - C,$$

c) $$K - C \text{ is open},$$

d) $$\text{bdry}\,(K - C) \subset C.$$

Assertion a) follows from (1), and b) is obvious. To prove c) we consider an arbitrary point s in $K - C$ and show that s is an interior point of $K - C$. According to Proposition 6 A) it is enough to demonstrate that for any neighborhood U of s in E_n satisfying $\overline{U}C = 0$ one can define a mapping of $K - C - U$ in S_{n-1} which cannot be extended over $K - C$. Because $K - U \neq K$, there exists by (2) an extension F of f over $C + (K - U) = (C + K) - U$. The partial mapping

$$F \mid K - C - U$$

cannot be extended over $K - C$, for if G were such an extension the mapping defined by $H(x) = G(x)$ for $x \in K - C$ and $H(x) = F(x)$ for $x \in C$ would yield an extension of f over $C + K$, contradicting (1). Having thus established *c*), *d*) follows immediately and completes the proof of the necessity.

Sufficiency. Suppose U is a non-empty open set contained in $X - C$ whose boundary B is contained in C; U is bounded as a subset of the compact set X. Let p be a point in U. Now $\pi_p \mid B$ cannot be extended over $U + B$ (Proposition 7 A)); since $B \subset C$ and $U + B \subset X$ it is a fortiori true that $\pi_p \mid C$ cannot be extended over X.

Theorem VI 13. *A compact subset C of E_n disconnects E_n if and only if there is an essential mapping of C in S_{n-1}.*

Proof. *Necessity.* For suppose C disconnects E_n. Then there certainly exist two points p and q separated by C. By Theorem VI 10 the mappings $\pi_p \mid C$ and $\pi_q \mid C$, of C in S_{n-1}, are in different homotopy classes. Hence at least one is essential.

Sufficiency. Let f be an essential mapping of C in S_{n-1}. Let I_n be an n-cube in E_n large enough to contain C. It is not possible to extend f over I_n because f would then be inessential (Example VI 7). By Theorem VI 12 therefore, replacing X by I_n, there is a non-empty subset U of $I_n - C$ which is open in E_n and whose boundary is contained in C. It is clear that U is a non-empty proper subset of $E_n - C$ which is both open and closed in $E_n - C$. Hence C disconnects E_n.

COROLLARY 1. *If a compact space C disconnects E_n then so does every subset of E_n which is homeomorphic to C.*

PROOF. For Theorem VI 13 has characterized disconnecting compact sets in E_n intrinsically.

COROLLARY 2. **Jordan Separation Theorem.** *A subset of E_n which is homeomorphic to S_{n-1} disconnects E_n.*

PROOF. This is an obvious application of Corollary 1.

COROLLARY 3. *Theorem* VI 13 *and Corollaries* 1 *and* 2 *hold when E_n is replaced by S_n.*

PROOF. It is easy to see that if C disconnects S_n it disconnects $S_n - p$ for any point p not in C, and conversely; and $S_n - p$ is homeomorphic to E_n.

An extremely elegant formulation of Theorem VI 13 can be made by making use of the concept of functional spaces:

Theorem VI 14.* *If C is a compact subset of E_n then $E_n - C$ is connected if and only if the functional space S_{n-1}^C is connected.*

PROOF. It is enough to observe that the components of the space S_{n-1}^C are the same as the homotopy classes of mappings of C in S_{n-1}, for on one hand every homotopy class is connected, even arcwise connected, by its very definition, and on the other hand each homotopy class is both open (as follows from Example VI 8) and closed.

* Theorems VI 13 and VI 14 are due to Borsuk: Über Schnitte der n-dimensionalen Euklidischen Räume, *Math. Ann.* 106 (1932), pp. 239–248. Theorem VI 4 is trivially true for $n = 1$ also.

CHAPTER VII

Dimension and Measure

This chapter is devoted to connections recently established* by Edward Szpilrajn between the concept of dimension and the concept of measure.

A p-dimensional measure for each non-negative real number p was defined by Hausdorff for arbitrary metric spaces. This measure is closely related to ordinary Lebesgue measure. It is a *metrical* concept, while dimension is purely *topological*. Nevertheless there is a strong connection between the two concepts, for it turns out (Theorem VII 2) that a space of dimension n must have positive n-dimensional measure. The converse, however, is not true (see Example VII 1). But if we consider not only the metric space X, but X together with all the metrics that can be put on it, or what is the same, the class of all spaces homeomorphic to X, then if all these spaces have positive n-dimensional measure, X itself must have dimension $\geqq n$. The fundamental result (proved by Theorems VII 2 and VII 4) is

Theorem VII 1. *A necessary and sufficient condition that a space X have dimension $\leqq n$ is that X be homeomorphic to a subset of I_{2n+1} whose $(n+1)$-dimensional measure is zero.*

Example VII 1. Both the set $\mathcal{J}$ of irrational points in the unit segment and the Cantor set $\mathcal{C}$ have dimension zero although $\mathcal{J}$ has (linear) measure unity and $\mathcal{C}$ has measure zero. But since† $\mathcal{J}$ is topologically contained in $\mathcal{C}$, there exists a topological image of $\mathcal{J}$ whose measure is zero.

1. p-measure in general metric spaces

Definition VII 1.‡ Let X be a space and p an arbitrary real number, $0 \leqq p < \infty$. Given $\epsilon > 0$, let§

* Szpilrajn: La dimension et la mesure, *Fund. Math.* 28 (1937), pp. 81–89. The proof of Theorem VII 5 is due to S. Eilenberg, loc. cit. p. 87.

† Each member of $\mathcal{J}$ has a unique representation as a sum $x = \sum_1^\infty a_n/2^n$ (scale of two). It is easy to see that $h(x) = \sum_1^\infty (2a_n)/3^n$ is a homeomorphism of $\mathcal{J}$ in $\mathcal{C}$.

‡ Hausdorff: Dimension und äusseres Mass, *Math. Ann.* 79 (1919), pp. 157–179, in particular p. 163.

§ $\delta(A)$, A an arbitrary subset of X, denotes the diameter of A; see index. We shall agree to set $[\delta(E)]^0 = 0$ if E is empty and $[\delta(E)]^0 = 1$ otherwise.

$$m_p^\epsilon = \inf \sum_{i=1}^{\infty} [\delta(A_i)]^p$$

where $X = A_1 + A_2 + \cdots$ is any decomposition of X in a countable number of subsets of diameter less than ϵ, and the superscript p denotes exponentiation. Let

$$m_p(X) = \sup_{\epsilon>0} m_p^\epsilon(X).$$

$m_p(X)$ is called the p- (or p-dimensional) measure of X.

The verification of the next three propositions is left to the reader.

A) It follows from footnote § on page 102 that
$m_0(X) = 0$ if X is empty,
$m_0(X) = n$ if X is a finite set of n points,
$m_0(X) = \infty$ if X is an infinite set.

B) If $p < q$ then $m_p(X) \geqq m_q(X)$; in fact $p < q$ and $m_p(X) < \infty$ imply $m_q(X) = 0$.

C) An n-dimensional polytope has finite n-measure. Consequently its q-measure is zero for all $q > n$.

D) A necessary and sufficient condition that a compact space C have p-measure zero is that there exist for each $\epsilon > 0$ a *finite* decomposition of C:

$$C = A_1 + \cdots + A_k$$

such that

(1) $$[\delta(A_1)]^p + \cdots + [\delta(A_k)]^p < \epsilon.$$

PROOF. The sufficiency of the condition is evident. We proceed to prove the necessity. Suppose then that $m_p(C) = 0$. By Definition VII 1 there exists a countable number of subsets A_1', A_2', $\cdots$ such that

$$C = A_1' + A_2' + \cdots$$

and

(2) $$\sum_{i=1}^{\infty} [\delta(A_i')]^p < \tfrac{1}{2}\epsilon.$$

It is possible to enlarge each A_i' slightly to an open set A_i so that

(3) $$[\delta(A_i)]^p < [\delta(A_i')]^p + \epsilon/2^{i+1}.$$

Since C is compact a finite number $A_1, \cdots, A_k$ of the A_i cover C. Formula (1) then follows from (2) and (3).

E) For subsets of the straight line 1-measure coincides with Lebesgue outer measure. However, the n-measure of a subset of E_n may differ numerically from its Lebesgue outer measure. Nevertheless the vanishing of the n-measure of a subset of E_n is equivalent to the vanishing of its Lebesgue outer measure.

2. An n-dimensional space has positive n-measure

Theorem VII 2. *Let X be a space of dimension n, $0 \leqq n < \infty$. Then* $m_n(X) > 0$.

This theorem is clearly equivalent (replacing n by $n + 1$) to

Theorem VII 3. *Let X be a space such that $m_{n+1}(X) = 0$, $(0 \leqq n < \infty)$. Then* $\dim X \leqq n$.

PROOF OF THEOREM VII 3. Let x_0 be an arbitrary point of X. For each $r > 0$ let $S(r)$ be the set of all points in X whose distance from x_0 is precisely r, i.e. the boundary of the spherical neighborhood of radius r around x_0. "For almost all r" will mean for all r except a set of Lebesgue measure zero.

A) If X is a space such that $m_{p+1}(X) = 0$, $0 \leqq p < \infty$, then for almost all r the sets $S(r)$ have p-measure zero.

Once this Proposition has been established Theorem VII 3 follows by induction,* since a set whose 0-measure is zero is empty (1 A)).

PROOF OF A). Let E be an arbitrary subset of X. Let

(1) $$r_1 = \inf_{x \in E} d(x, x_0), \qquad r_2 = \sup_{x \in E} d(x, x_0).$$

It is clear that

(2) $$r_2 - r_1 \leqq \delta(E)$$

From (1) and (2) it follows that (the integrals being taken as upper integrals)

(3) $$\int_0^\infty [\delta(S(r)\cdot E)]^p dr = \int_{r_1}^{r_2} [\delta(S(r)\cdot E)]^p dr$$
$$\leqq [\delta(E)]^p \int_{r_1}^{r_2} dr \leqq [\delta(E)]^{p+1}.$$

* Actually Proposition A) is stronger than Theorem VII 3. For from A) it clearly follows that if $m_{n+1}(X) = 0$ then not only does every point $x \in X$ have

We are given that $m_{p+1}(X) = 0$. From Definition VII 1 we derive the existence of a sequence of decompositions of X:

$$X = A_1^n + A_2^n + \cdots$$

such that

$$\lim_{n\to\infty} \sum_{i=1}^{\infty} [\delta(A_i^n)]^{p+1} = 0. \tag{4}$$

From (3) and (4) we get

$$\lim_{n\to\infty} \sum_{i=1}^{\infty} \int_0^{\infty} [\delta(S(r)\cdot A_i^n)]^p dr = 0.$$

Each of the integrands is non-negative. Hence we may interchange integration and summation. This gives us

$$\lim_{n\to\infty} \int_0^{\infty} \sum_{i=1}^{\infty} [\delta(S(r)\cdot A_i^n)]^p dr = 0;$$

that is*

$$\underset{n\to\infty}{\text{l.i.m.}} \sum_{i=1}^{\infty} [\delta(S(r)\cdot A_i^n)]^p = 0.$$

Hence† there exists a subsequence $A_i^{n_k}$ of the decompositions of X such that

$$\lim_{k\to\infty} \sum_{i=1}^{\infty} [\delta(S(r)\cdot A_i^{n_k})]^p = 0 \text{ for almost all } r.$$

This implies

$$m_p(S(r)) = 0 \text{ for almost all } r,$$

which proves A) and hence Theorem VII 3.

3. An n-dimensional space is homeomorphic to a space of $(n+1)$-measure zero

Theorem VII 4. *If a space has dimension $\leqq n$ it is homeomorphic to a subset of I_{2n+1} of $(n+1)$-measure zero.*

---

arbitrarily small neighborhoods with boundaries of dimension $\leqq n - 1$ but *almost all* of the *spherical* neighborhoods of x have boundaries of dimension $\leqq n - 1$.

* "l.i.m." stands for "limit in the mean."

† E. C. Titchmarsh, *The Theory of Functions*, Oxford, 1932, section 12.5, in particular p. 388.

Theorem VII 4 is contained in

Theorem VII 5. *If a space X has dimension $\leqq n$ then there is a homeomorphism h of X in I_{2n+1} such that for every real number $r > n$*

$$m_r(\overline{h(X)}) = 0. \tag{1}$$

Moreover, the space $I_{2n+1}{}^X$ contains a dense G_δ-set of homeomorphisms satisfying (1).

PROOF OF THEOREM VII 5. Suppose $q > n$. Consider the class K_q of all $f \in I_{2n+1}{}^X$ (see Chapter V) such that

$$m_q(\overline{f(X)}) = 0.$$

Now $\overline{f(X)}$ is compact, as a closed subset of I_{2n+1}. Hence (1 D)) $f \in K_q$ means that for every integer $i = 1, 2, \cdots$ there exists a finite decomposition, denoted by (d),

$$(d): \qquad X = A_1 + \cdots + A_k$$

for which

$$[\delta(f(A_1))]^q + \cdots + [\delta(f(A_k))]^q < 1/i. \tag{2}$$

In symbols:

$$K_q = \prod_{i=1}^{\infty} \sum_{(d)} G_{i,q}^{(d)},$$

the summation running over all finite decompositions (d), and $G_{i,q}^{(d)}$ denoting the class of all f satisfying (2). But $G_{i,q}^{(d)}$ is clearly open; hence

$$K_q \text{ is a } G_\delta \text{ in } I_{2n+1}{}^X.$$

Let K^* be the class of all $f \in I_{2n+1}{}^X$ such that $\overline{f(X)}$ is contained in an n-dimensional polytope. K^* is dense by Remark 1 on page 70; and because $q > n$ we have (1 C)) that $K^* \subset K_q$. Hence

$$K_q \text{ is a dense } G_\delta \text{ in } I_{2n+1}{}^X.$$

Let H be the dense G_δ of homeomorphisms of X in I_{2n+1} given by Theorem V 3. Let

$$H^* = H \prod_{i=1}^{\infty} K_{n+(1/i)}. \tag{5}$$

Then H^* is a dense G_δ in $I_{2n+1}{}^X$, by Baire's Theorem, and hence is not empty. Moreover $h \in H^*$ implies

a) h is a homeomorphism of X in $\overline{I_{2n+1}}$,

b) $m_r(h(X)) = 0$ for every $r > n$.

This proves the theorem.

REMARK 1. Theorems VII 5 and VII 3 prove anew that a space can be imbedded in a compact space of the same dimension (Theorem V 6).

REMARK 2. A consequence of Theorem VII 4 and Proposition 2 A) is the following. Let X be a space of dimension $\leqq n$. Then there is a remetrization X' of X with this property: not only does every point $x \in X'$ have arbitrarily small neighborhoods with boundaries of dimension $\leqq n - 1$ but *almost all* of the *spherical* neighborhoods of x have boundaries of dimension $\leqq n - 1$.

4. Hausdorff dimension

Given an arbitrary metric space X we denote by the Hausdorff dimension* of X the supremum of all real numbers p such that $m_p(X) > 0$. It follows from Theorem VII 2 that

$$\text{Hausdorff dim } X \geqq \dim X.$$

The Hausdorff dimension of a space need not be an integer. Thus the Hausdorff dimension of the Cantor set is† $\log 2/\log 3 = 0.63093$. But it follows from Theorem VII 5 that if we let X' range over all the spaces homeomorphic to a given space X, then

$$\inf \text{Hausdorff dim } X' = \dim X.$$

* Hausdorff: Dimension und äusseres Mass, loc. cit., p. 166.

† Hausdorff: Dimension und äusseres Mass, loc. cit., p. 172.

CHAPTER VIII

Homology and Dimension

The beginning chapters of this book were dominated by point-set theoretical methods. Combinatorial methods made their first appearance in Chapter IV (in proving that Euclidean n-space has dimension n), and gained more prominence in Chapter V, where the proof of the Imbedding Theorem was obtained by combining the point-set topology of functional spaces with the combinatorial topology of complexes. In this chapter we investigate dimension theory from a purely algebraic, combinatorial, point of view. The link connecting this new method with our previous discussion is Theorem VI 4, on mappings in spheres. The main goal of this chapter is the characterization of dimension by algebraic properties (Theorem VIII 3) due to Alexandroff.†

A large part of this chapter is concerned with a condensed exposition of algebraic connectivity theory.* Although this exposition is not directly connected with the essential content of the book, it has seemed desirable to the authors to include it because of the rapidly changing aspect of connectivity theory. If one were willing to operate only with cohomology, omitting the discussion of the duality relations between homology and cohomology, the volume of this exposition would be very substantially reduced.

Throughout this chapter the word group means commutative group, with the group operation written as addition. The two most frequently used are the group of the integers and the group of the real numbers modulo 1. We denote these by $\mathfrak{J}$ and $\mathfrak{R}_1$ respectively. $\mathfrak{R}_1$ is, of course, isomorphic to the group of rotations of the circumference and to the multiplicative group of complex numbers of modulus 1.

1. Combinatorial connectivity theory of a complex

Definition VIII 1. Let K be a (finite) complex (see page 67) and n

† Dimensionstheorie, *Math. Ann.* 106 (1932), pp. 161–238.

* Some acquaintance with combinatorial topology and group theory is assumed in this chapter although the chapter is logically quite self-contained and all the basic definitions are given explicitly. We recommend Chapters 2 and 3 of Seifert-Threlfall, *Lehrbuch der Topologie*, B. G. Teubner, Leipzig, 1934, as an excellent introduction to combinatorial topology. Pontrjagin's *Topological Groups*, Princeton University Press, 1939, (cited as Pontrjagin), in particular Chapters 1, 3, 5, contains complete discussions of the required group theory.

an integer, $1 \leqq n$. An *oriented* n-simplex s_n of K is an n-simplex of K whose vertices are written in a definite order:

$$s_n = (p_0, p_1, \cdots, p_n).$$

Oriented simplexes made up of the same vertices ordered in different ways are considered as *equal* if they can be obtained from each other by an even permutation. Hence to each non-oriented n-simplex correspond two different oriented simplexes; if one of them is denoted by the symbol s or $+ s$, then the other (obtained from s by an odd permutation of its vertices) will be denoted by $- s$. For example

$$(p_0, p_1, p_2) = (p_1, p_2, p_0) = (p_2, p_0, p_1) = -(p_0, p_2, p_1)$$
$$= -(p_2, p_1, p_0) = -(p_1, p_0, p_2).$$

We shall find it convenient to extend the definitions of oriented n-simplexes to the cases $n = 0$ and $n = -1$, and this we do as follows: With each non-oriented 0-simplex of K, i.e. with each vertex p of K, we associate two symbols, $+ p$ and $- p$, and we call these the *oriented* 0-simplexes belonging to p. Further we agree to call the empty set of vertices of K the* non-oriented (-1)-simplex of K and to associate with it two symbols $+ \Lambda$ and $- \Lambda$ as the *oriented* (-1)-*simplexes of* K. Of course we set

$$-(+p) = -p, \; -(-p) = +p, \; -(+\Lambda) = -\Lambda, \; -(-\Lambda) = +\Lambda.$$

An oriented n-simplex $(p_0, p_1, \cdots, p_n)$, $n \geqq 1$, is called an *oriented face of an oriented* $n + 1$-*simplex* s_{n+1} if

$$s_{n+1} = (p_{n+1}, p_0, \cdots, p_n).$$

For $n \geqq 2$ any oriented n-simplex has $n + 1$ oriented faces. Thus the oriented 2-simplex (p_0, p_1, p_2) has the three oriented faces

$$(p_1, p_2) = -(p_2, p_1), \quad (p_2, p_0) = -(p_0, p_2), \quad (p_0, p_1) = -(p_1, p_0).$$

We define the *oriented faces of a* 1-*simplex* (p_0, p_1) as the two oriented 0-simplexes p_1 and $- p_0$; and we agree to regard $+ \Lambda$ as the only *oriented face of* 0-*simplexes* of the type $+ p$ and $- \Lambda$ as the only *oriented face of* 0-*simplexes* of the type $- p$.

In the following paragraphs the symbol s_n will be used to denote an arbitrary oriented n-simplex of K, $-1 \leqq n$, with superscripts to single out individual simplexes. We write $s_n \prec s_{n+1}$ or $s_{n+1} \succ s_n$ if s_n is a face of s_{n+1}, and we write $s_n \prec s_{n+2}$ or $s_{n+2} \succ s_n$ if an s_{n+1} exists such that

* The empty set of vertices is the only (-1)-simplex of K.

$s_n \prec s_{n+1} \prec s_{n+2}$. The reader will easily verify the following simple facts.

A) The relation $s_n \prec s_{n+1}$ implies $-s_n \prec -s_{n+1}$, but excludes $s_n \prec -s_{n+1}$ and $-s_n \prec s_{n+1}$. The relations

$$s_n \prec s_{n+2}, \qquad s_n \prec -s_{n+2}, \qquad -s_n \prec s_{n+2}, \qquad -s_n \prec -s_{n+2}$$

are equivalent, and assert that all the vertices of s_n are vertices of s_{n+2}, i.e. that the non-oriented simplex s_n is a face of the non-oriented simplex s_{n+2}. If $s_n \prec s_{n+2}$ there exists only one oriented simplex s_{n+1} such that $s_n \prec s_{n+1} \prec s_{n+2}$.

Chains, cycles, cocycles

Definition VIII 2. Let G be an abelian group written in additive form. Let K be a complex and n an integer. By an *(n, G)-chain of K*, or an *n-chain of K with basis G*, we mean simply a function φ associating an element of G with each oriented n-simplex of K, and satisfying

$$\varphi(-s_n) = -\varphi(s_n).$$

The zero (n, G)-chain of K is the function associating the element 0 of G with each oriented n-simplex of K.

For each $n > \dim K$ the one and only (n, G)-chain of K is the zero (n, G)-chain.

Remark. A $(-1, G)$-chain can be identified with the element $\varphi(+\Lambda)$ of G. A $(0, G)$-chain is essentially a function of the vertices p with values in G. For $n \geqq 1$ an (n, G)-chain can be regarded as a skew-symmetric function of $n + 1$ vertices with values in G (this function is defined only for those sets of $n + 1$ vertices which determine an n-simplex of K). If G is the group of integers mod 2 an (n, G)-chain can be regarded as a collection of non-oriented n-simplexes (namely the collection of simplexes satisfying the relation $\varphi(s_n) = \varphi(-s_n) \neq 0$).

Definition VIII 3. By the *sum*, $\varphi = \varphi_1 + \varphi_2$, of two (n, G)-chains φ_1 and φ_2 of K we mean the (n, G)-chain φ defined by the relation

$$\varphi(s_n) = \varphi_1(s_n) + \varphi_2(s_n).$$

With this definition of sum the (n, G)-chains of K form an additive group, called the *group of (n, G)-chains of K* and denoted by $C_n(K, G)$, or, when no confusion is possible, by $C_n(K)$ or even C_n. The zero of this group is, of course, the zero (n, G)-chain.

If $n > \dim K$ then $C_n(K, G)$ reduces to the single element 0.

Given a simplex $s_n^0 \varepsilon K$ and an element $g_0 \varepsilon G$, we denote by the symbol $g_0 s_n^0$ the n-chain φ defined as follows:

$$\begin{cases} \varphi(s_n) = g_0 & \text{if} \quad s_n = s_n^0, \\ \varphi(s_n) = -g_0 & \text{if} \quad s_n = -s_n^0, \\ \varphi(s_n) = 0 & \text{otherwise.} \end{cases}$$

Chains of this type will be called *elementary* chains. Each (n, G)-chain can be represented as a sum $\sum_i g_i s_n^i$ of elementary (n, G)-chains,* this representation being unique provided we neglect terms with coefficient zero and require that $s_n^i \neq \pm s_n^{i'}$ for $i \neq i'$.

Definition VIII 4. The *boundary*, $\beta\varphi$, of an (n, G)-chain $\varphi(n \geqq 0)$ is the $(n - 1, G)$-chain defined by the formula

$$\beta\varphi(s_{n-1}) = \sum_{s_n > s_{n-1}} \varphi(s_n).$$

The *coboundary*,† $\gamma\varphi$, of an (n, G)-chain φ $(n \geqq -1)$ is the $(n + 1, G)$-chain defined by the formula‡

$$\gamma\varphi(s_{n+1}) = \sum_{s_n < s_{n+1}} \varphi(s_n).$$

Example VIII 1. The boundary of the elementary chain gs_n^0 $(n \geqq 0)$ is the chain

$$\sum_{s_{n-1} < s_n^0} gs_{n-1},$$

and the coboundary of gs_n^0 is the chain

$$\sum_{s_{n+1} > s_n^0} gs_{n+1}.$$

B) The operator β is obviously a homomorphism of C_n in C_{n-1}, and γ is a homomorphism of C_n in C_{n+1}. The operators β and γ have the important properties

$$\beta\beta\varphi = 0, \qquad \gamma\gamma\varphi = 0,$$

for any (n, G)-chain φ (with the restriction $n \geqq 1$ for the first relation).

* This is essentially the more familiar procedure of defining a chain as a symbolic linear form of simplexes. It shows that $C_n(K, G)$ is the direct sum of k groups isomorphic to G, k being the number of unoriented n-simplexes in K.

† The concept of coboundary is fairly recent, while classical combinatorial topology employed only the boundary concept.

‡ We regard a sum extended over an empty collection as equal to zero.

PROOF. By the definitions of β and γ and the last part of Proposition A) we have

$$\beta\beta\varphi(s_{n-2}) = \sum_{s_n > s_{n-2}} \varphi(s_n), \qquad \gamma\gamma\varphi(s_{n+2}) = \sum_{s_n < s_{n+2}} \varphi(s_n).$$

Also from A) we know that $s_n < s_{n+2}$ implies $s_n < -s_{n+2}$ and $-s_n < s_{n+2}$. Hence each term $\varphi(s_n)$ under the $\sum$-sign is cancelled by its negative $\varphi(-s_n) = -\varphi(s_n)$, and this shows that $\beta\beta\varphi(s_{n-2}) = 0$ for each s_{n-2} and $\gamma\gamma\varphi(s_{n+2}) = 0$ for each s_{n+2}.

Definition VIII 5. An (n, G)-chain is called a *cycle* if its boundary is equal to zero, a *cocycle* if its coboundary is equal to zero. It follows from B) that any (n, G)-chain which is the boundary of an $(n + 1, G)$-chain (the coboundary of an $(n - 1, G)$-chain) is a cycle (cocycle). Chains of this type are called *bounding cycles* (*cobounding cocycles*).

Both the (n, G)-cycles of K and the (n, G)-cocycles of K clearly form subgroups* of $C_n(K, G)$, and we denote these by $Z_n(K, G)$ and $Z^n(K, G)$ respectively. Further, the (n, G)-bounding cycles form a subgroup† of $Z_n(K, G)$, by B), which we denote by $B_n(K, G)$. Similarly the (n, G)-cobounding cocycles form a subgroup† of $Z^n(K, G)$, and we denote this by $B^n(K, G)$.

HOMOLOGY AND COHOMOLOGY GROUP OF COMPLEXES

Definition VIII 6. The difference group $Z_n(K, G) - B_n(K, G)$, of n-cycles by bounding n-cycles, is called the *n-dimensional homology group of K with basis G* and is denoted by $H_n(K, G)$ or $H_n(K)$ or H_n. The elements of H_n, i.e. the cosets of B_n in Z_n, are called *n-dimensional homology classes*, and two cycles belonging to the same homology class are said to be *homologous*. Similarly the difference group $Z^n(K, G) - B^n(K, G)$, of the n-cocycles by the cobounding n-cocycles, is called the *n-dimensional cohomology group of K with basis G* and is denoted by $H^n(K, G)$ or $H^n(K)$ or H^n. The elements of H^n, i.e. the cosets of B^n in Z^n, are called *n-dimensional cohomology classes*, and two cocycles belonging to the same cohomology class are said to be *cohomologous*.

It is clear that two n-cycles φ and ψ are homologous if and only if there exists an $(n + 1)$-chain θ such that $\beta\theta = \varphi - \psi$, and two

* The *kernel* of a homomorphism h of a group G in a group G' is the set of elements g of G such that $h(g) = 0$. The subgroups Z_n and Z^n are thus the kernels of the homomorphisms β and γ of C_n in C_{n-1} and C_n in C_{n+1}.

† B_n is the image of C_{n+1} under the homomorphism β and B^n is the image of C_{n-1} under the homomorphism γ.

n-cocycles φ and ψ are cohomologous if and only if there exists an $(n-1)$-chain θ such that $\gamma\theta = \varphi - \psi$.

REMARK. Let $\mathfrak{J}$ denote the group of integers. Then for each n, $H_n(K, \mathfrak{J})$ is a group with a finite number of generators. It can therefore* be split into the direct sum of a free group and a certain number of groups of finite orders. The rank of the free subgroup is called the nth *Betti number* of K, and the orders of the finite subgroups, in canonical decomposition, are called the *n-dimensional torsion coefficients of K*.

It is known that the Betti numbers and torsion coefficients of a complex completely determine the homology and cohomology groups of the complex with arbitrary basis group G. For example $H^n(K, \mathfrak{J})$ is the direct sum of b_n infinite cyclic groups, b_n being the nth Betti number of K, and of finite cyclic groups whose orders are the $(n-1)$-dimensional torsion numbers of K. $H_n(K, \mathfrak{R}_1)$ is the direct sum of b_n groups isomorphic to $\mathfrak{R}_1$ and of finite cyclic groups whose orders are the $(n-1)$-dimensional torsion numbers of K.

EXAMPLE VIII 2. If $n > \dim K$, then $H^n(K, G) = H_n(K, G) = 0$.

EXAMPLE VIII 3. A complex K is called *connected* if it cannot be split into two complexes without common vertices or, what amounts to the same, if any two vertices p and q of K can be joined by a sequence of 1-simplexes $(p, p_1), (p_1, p_2), \cdots, (p_n, q)$. Every complex K can be decomposed in a unique way into connected subcomplexes called *components* of K. The reader will easily verify that a $(0, G)$-chain of K, i.e. a function $\varphi(p)$ of the vertices p, is a cocycle if and only if, given any component of K, $\varphi(p)$ has a constant value on it. φ is cohomologous to 0 if and only if φ is a constant on all of K. Consequently the group $H^0(K, G)$ is the direct sum of $m - 1$ groups each isomorphic to G, where m denotes the number of components of K. The same result is easily established for the group $H_0(K, G)$.

EXAMPLE VIII 4. If $n = \dim K$ the group $H_n(K, G)$ coincides with the group $Z_n(K, G)$. Moreover every n-chain is a cocycle and hence $H^n(K, G)$ coincides with the group $C^n(K, G) - B^n(K, G)$.

EXAMPLE VIII 5. Let K be an n-dimensional complex with the following properties: 1) each non-oriented $(n-1)$-simplex of K is a common face of exactly two n-simplexes; 2) K cannot be represented as a sum of two complexes which have no $(n-1)$-simplex in common. (Complexes of this kind are sometimes called *n-dimensional pseudomanifolds.*) We take for G the group $\mathfrak{J}$ and wish to determine the cohomology group $H^n(K) = H^n(K, \mathfrak{J})$. Each n-chain is a cocycle

* Seifert-Threlfall, *Lehrbuch der Topologie*, B. G. Teubner, Leipzig, 1934, §86, in particular Satz III, p. 308.

and the only question we have to answer is: when is an n-chain cobounding? Two cases are to be distinguished:

a) Suppose one can assign to each n-simplex of K a definite orientation, called the *positive* orientation, in such a way that every oriented $(n - 1)$-simplex is an oriented face of exactly one positively oriented n-simplex (and consequently of exactly one negatively oriented simplex). We say then that the complex K is *orientable.* The reader will easily prove that given any two positively oriented simplexes s_n and $\tilde{s}_n$ the elementary chains $1s_n$ and $1\tilde{s}_n$ are cohomologous (this is clear in case s_n and $\tilde{s}_n$ are adjacent simplexes; the general statement is reduced to property 2)). It will further be observed that a cobounding n-chain satisfies the condition $\sum \varphi(s_n) = 0$, where the sum is extended over all positively oriented simplexes (it is sufficient to verify this for the case that φ is the coboundary of an elementary chain). From this it can be concluded that the group $H^n(K)$ is isomorphic to $\mathfrak{J}$ and that for a given oriented simplex s_n^0 the elementary chains ms_n^0, $m = 0, \pm 1, \pm 2, \cdots$, represent all the cohomology classes, once each. Given an arbitrary group G, the group $H^n(K, G)$ is, similarly, isomorphic to G.

b) Suppose it is not possible to orient the n-simplexes of K in accordance with the requirements above. K is said in this case to be *non-orientable* and it is easy to show that for each s_n the chain $2\, s_n$ is cohomologous to 0 and the group $H^n(K, \mathfrak{J})$ is of order 2. The non-zero element of $H^n(K, \mathfrak{J})$ is represented by each of the cocycles $1\, s_n$. (In order to prove that $1\, s_n$ is not cohomologous to 0 let $\pm\, s_n^i$, $i = 1, \cdots, k$, be all the oriented n-simplexes of K and show that if φ is a cobounding n-chain then $\sum_{i=1}^k \varphi(s_n^i) \equiv 0 \bmod 2$).

EXAMPLE VIII 6. Let the complex K be the sum of two complexes K_1 and K_2 without common vertices. Then, provided $n > 0$, $H^n(K, G)$ is the direct sum of the groups $H^n(K_1, G)$ and $H^n(K_2, G)$ and $H_n(K, G)$ is the direct sum of $H_n(K_1, G)$ and $H_n(K_2, G)$. For $n = 0$ these statements are false (see Example VIII 3).

We shall often consider correspondences between the chains of one complex and the chains of another.

C) Let K_1 and K_2 be two complexes and G a group. Suppose to each (n, G)-chain φ of K_1 is assigned an (n, G)-chain $h(\varphi)$ of K_2, $n = 0, 1, 2, \cdots$, and for each n the operator h is a homomorphism of $C_n(K_1, G)$ in $C_n(K_2, G)$. Suppose finally that h commutes with the boundary operator β, i.e.

$$h(\beta(\varphi)) = \beta(h(\varphi))$$

for every n-chain φ of K_1. Then h sends cycles into cycles and bounding cycles into bounding cycles, and consequently maps $H_n(K_1)$ homomorphically in $H_n(K_2)$. Similarly, if h commutes with the coboundary operator γ, i.e. if

$$h(\gamma(\varphi)) = \gamma(h(\varphi)),$$

then h sends cocycles into cocycles and cobounding cocycles into cobounding cocycles, and consequently maps $H^n(K_1)$ homomorphically in $H^n(K_2)$.

Natural homomorphisms. Homologies and cohomologies mod L

Let K be a complex, L a subcomplex* of K. To each (n, G)-chain Φ of K corresponds in a natural way the (n, G)-chain $\varphi = h_L\Phi$ of L obtained by restricting the range of the variable in Φ to simplexes of L, and conversely to each chain φ of L corresponds the chain $\Phi = h_K\varphi$ of K defined by $\Phi(s_n) = \varphi(s_n)$ if $s_n \varepsilon L$ and $\Phi(s_n) = 0$ otherwise. For each n we obtain this way a homomorphism h_L of the group C_n (K, G) in† the group $C_n(L, G)$ and a homomorphism h_K of the group $C_n(L, G)$ in† the group $C_n(K, G)$. It is easy to show that the operator h_L commutes with the coboundary operator γ, i.e. for any chain Φ of K,

(1) $$h_L\gamma\Phi = \gamma h_L\Phi.$$

Similarly the operator h_K commutes with the boundary operator β, i.e. for any chain φ of L,

(1′) $$h_K\beta\varphi = \beta h_K\varphi.$$

From C) we conclude that h_L induces a homomorphism of $H^n(K, G)$ in $H^n(L, G)$ and h_K induces a homomorphism of $H_n(L, G)$ in $H_n(K, G)$.

Definition VIII 7. The homomorphism h_L of $H^n(K, G)$ in $H^n(L, G)$

* It is essential to distinguish between the chains of K and the chains of L: the former are functions with range consisting of the simplexes of K and the latter are functions with range consisting of the simplexes of L. Even if (for a fixed n) the set of n-simplexes of L coincides with the set of n-simplexes of K it is necessary to distinguish between an n-chain Φ of K and an n-chain φ of L: The range of the function $\gamma\Phi$ will consist of all the $(n + 1)$-simplexes of K, while the range of $\gamma\varphi$ will consist of the $(n + 1)$-simplexes of L only. One must be especially careful when writing a chain as a linear combination of elementary chains, for an elementary chain $g\ s_n$, where s_n^0 is a simplex of L, might mean either a chain of L or a chain of K.

† h_L is actually a homomorphism of $C_n(K, G)$ *on* $C_n(L, G)$ and h_K is actually an *iso*morphism of $C_n(L, G)$ in $C_n(K, G)$. (In this book both homomorphism and isomorphism of a group G *in* a group H mean mappings of G on H or subgroups of H.

and the homomorphism h_K of $H_n(L, G)$ in $H_n(K, G)$ are called the *natural homomorphisms.* If the element e of $H^n(L)$ is the image of an element $\tilde{e}$ of $H^n(K)$ under h_L we call $\tilde{e}$ an *extension* of e and we call the element e *extendable over* K. An element of $H_n(L)$ which is sent into the zero of $H_n(K)$ under h_K is said to *bound in* K.

We omit the simple proof of the following Proposition:

D) Let the dimension of K be n. The natural homomorphism of $H_n(L)$ in $H_n(K)$ is an *iso*morphism of $H_n(L)$ in $H_n(K)$, while the natural homomorphism of $H^n(K)$ in $H^n(L)$ is a homomorphism of $H^n(K)$ *on* $H^n(L)$, i.e. every element of $H^n(L)$ has an extension (use Example VIII 4). Suppose L is the subcomplex of K made up of all the simplexes of K of dimension $\leqq m$. For $m = n - 1$ the natural homomorphism of $H_m(L)$ in $H_m(K)$ is a homomorphism of $H_m(L)$ *on* $H_m(K)$, while the natural homomorphism of $H^m(K)$ in $H^m(L)$ is an *iso*morphism of $H^m(K)$ in $H^m(L)$. If $m \leqq n - 2$ both natural homomorphisms are *iso*morphisms *on*.

E) Let e be an element of $H^n(L)$ and $\tilde{e} \in H^n(K)$ an extension of e. Then to every cocycle φ of L which represents e there exists a cocycle Φ of K which represents $\tilde{e}$ and is an extension of φ.

PROOF. First suppose $\tilde{e} = 0$. Then $e = 0$, i.e. there exists a cocycle ψ of L such that $\gamma\psi = \varphi$. The cobounding cocycle $\gamma h_K \psi$ is then an extension of φ, for $h_L \gamma h_K \psi = \gamma h_L h_K \psi = \gamma\psi = \varphi$. Now let $\tilde{e}$ be arbitrary and Φ_1 a cocycle of K in the cohomology class $\tilde{e}$. Then $\varphi - h_L \Phi_1$ is a cobounding cocycle of L and hence extendable to a cobounding cocycle Φ_2 of K. The cocycle $\Phi_1 + \Phi_2$ is the desired extension of φ.

We now consider functions with values in G defined for the oriented n-simplexes* of $K - L$. If such a function φ satisfies the condition $\varphi(-s_n) = -\varphi(s_n)$ it is called an (n, G)-*chain modulo* L. The (n, G)-chains mod L form a group. The boundary and coboundary of a chain φ mod L are defined in the same way as for ordinary chains with the only difference that $\beta\varphi$ and $\gamma\varphi$ are to be regarded as defined only for $s_n \in K - L$. We verify that $\beta\beta\varphi = 0$ and $\gamma\gamma\varphi = 0$. As before we define cycles and cocycles mod L by the relations $\beta\varphi = 0$ and $\gamma\varphi = 0$. We define $H_n(K \text{ mod } L, G)$, the nth *homology group of* K mod L, as the difference group of the cycles mod L by the bounding cycles mod L and $H^n(K \text{ mod } L, G)$, the nth *cohomology* group of K mod L, as the difference group of the cocycles mod L by the cobounding cocycles mod L.

* This means n-simplexes of K not belonging to L. K is regarded as the set of its simplexes.

EXAMPLE VIII 8. Let K be an arbitrary complex. We form a new complex K^*, called the *cone* over K, by adding to the vertices of K a new vertex p and defining the simplexes of K^* as the simplexes of K and in addition the simplexes (p, s_n), where s_n is a simplex of K. It is clear that the oriented n-simplexes of $K^* - K$ are in 1:1 correspondence with the oriented $(n-1)$-simplexes of K; hence the (n, G)-chains of K^* mod K are in isomorphic correspondence with the $(n-1, G)$-chains of K. The reader will prove without difficulty that the groups $H^n(K^* \bmod K)$ and $H^{n-1}(K)$ are isomorphic, and so also are $H_n(K^* \bmod K)$ and $H_{n-1}(K)$, $n=1, 2, \cdots$.

EXAMPLE VIII 9. If K_1, K_2, L are subcomplexes of K and $K = K_1 + K_2$, $K_1K_2 \subset L$, the group $H^n(K \bmod L, G)$ is the direct sum of the groups $H^n(K_1 \bmod LK_1, G)$ and $H^n(K_2 \bmod LK_2, G)$. An analogous statement holds for homology groups. (In contrast with Example VIII 6 these statements hold even for $n = 0$.)

If Φ is an m-chain of K we denote by $h_{K-L}\Phi$ the m-chain mod L obtained from Φ by restricting the range of the variable to the simplexes of $K - L$. If φ is a chain mod L we denote by $h_K\varphi$ the m-chain of K defined by $h_K\varphi(s_n) = \varphi(s_n)$ if $s_n \in K - L$ and $h_K\varphi(s_n) = 0$ if $s_n \in L$. One verifies easily that

$$h_{K-L}\beta\Phi = \beta h_{K-L}\Phi \tag{2}$$

for any m-chain Φ of K, and

$$h_K\gamma\varphi = \gamma h_K\varphi \tag{2'}$$

for any m-chain φ of K mod L. (Compare these formulas with the formulas (1) and (1′) above† and note the interchange of the operators β and γ.)

F) Suppose $H^m(K \bmod L, G) = 0$. Let Φ be an (m, G)-cocycle of K and ψ an $(m-1, G)$-chain of L such that $\gamma\psi = h_L\Phi$. Then there exists an $(m-1)$-chain Ψ of K with the properties: $h_L\Psi = \psi$; $\gamma\Psi = \Phi$.

PROOF. $\Phi' = \Phi - \gamma h_K\psi$ is an m-cocycle of K with the property $h_L\Phi' = 0$, i.e. $\Phi'(s_m) = 0$ if $s_m \in L$, for by (2′), $h_L\Phi' = h_L\Phi - \gamma\psi = 0$. Obviously $h_{K-L}\Phi'$ is an m-cocycle mod L, and since by hypothesis all the m-cocycles mod L are cobounding, there exists an $(m-1)$-chain ψ' mod L such that $\gamma\psi' = h_{K-L}\Phi'$ and hence, by (2′), $\gamma h_K\psi' = h_K\gamma\psi'$

† Relations (2) and (2′) show that the operator k_K induces, for a given m and G, a homomorphism of the group H^m $(K \bmod L, G)$ in the group H^m (K, G), and h_{K-L} induces a homomorphism of the group H_m (K, G) in the group H_m $(K \bmod L, G)$.

$= h_K h_{K-L} \Phi' = \Phi' = \Phi - \gamma h_K \psi$. The chain $\Psi = h_K \psi + h_K \psi'$ satisfies the conditions $h_L \Psi = \psi$ and $\gamma \Psi = \Phi$.

From F) we derive:

G) If $H^m(K \bmod L, G) = 0$ the natural homomorphism of $H^m(K, G)$ in $H^m(L, G)$ is an *iso*morphism, and the natural homomorphism of $H^{m-1}(K, G)$ in $H^{m-1}(L, G)$ is a homomorphism *on* $H^{m-1}(L, G)$.

PROOF. On one hand F) implies that an (m, G)-cocycle Φ of K cobounds if $h_L \Phi$ cobounds. On the other hand F) applied to the m-cocycle $\Phi = 0$ asserts that every $(m - 1, G)$-cocycle ψ of L is the image of a cocycle of K under h_L.

We now establish for the boundary operator β propositions similar to F) and G). For the sake of abbreviation we shall call two (m, G)-chains Φ_1 and Φ_2 of K (which need not be cycles) *homologous* if their difference $\Phi_1 - \Phi_2$ is a bounding cycle; this implies of course that Φ_1 and Φ_2 have the same boundary.

F′) Suppose $H_m(K \bmod L, G) = 0$. Let φ be an $(m - 1, G)$-cycle of L and Ψ an (m, G)-chain of K such that $\beta\Psi = h_K \varphi$. Then there exists an (m, G)-chain ψ of L such that $h_K \psi$ is homologous to Ψ. This implies $\beta\psi = \varphi$ (for h_K is a 1:1 correspondence, and by (1′), $h_K \beta \psi = \beta h_K \psi = \beta\Psi = h_K \varphi$).

PROOF. $h_{K-L}\Psi$ is an m-cycle mod L (for by (2), $\beta h_{K-L}\Psi = h_{K-L}\beta\Psi = h_{K-L} h_K \varphi = 0$), and since by hypothesis all (m, G)-cycles mod L bound, $h_{K-L}\Psi = \beta\theta$, where θ is an $(m + 1)$-chain mod L. The m-chain $\Psi' = \Psi - \beta h_K \theta$ is homologous to Ψ, and by (2) we have $h_{K-L}\Psi' = h_{K-L}\Psi - \beta h_{K-L} h_K \theta = h_{K-L}\Psi - \beta\theta = 0$; this shows that $\Psi' = h_K h_L \Psi'$, and consequently the chain $\psi = h_L \Psi'$ satisfies the condition of F′).

From F′) we derive:

G′) If $H_m(K \bmod L, G) = 0$ the natural homomorphism of $H_m(L, G)$ in $H_m(K, G)$ is a homomorphism *on* $H_m(K, G)$, and the natural homomorphism of $H_{m-1}(L, G)$ in $H_{m-1}(K, G)$ is an *iso*morphism.

PROOF. F′) obviously implies that an $(m - 1, G)$-cycle φ of L bounds if $h_K \varphi$ bounds. On the other hand F′) applied to the $(m - 1)$-cycle $\varphi = 0$ asserts that every (m, G)-cycle of K is the image of an m-cycle of L under h_K.

We use Propositions G) and G′) to determine the homology and cohomology groups of the simplest complexes.

Example VIII 10. Let Q_n be the complex made up of a single n-simplex and its faces. Then for all m, the cohomology and homology groups $H^m(Q_n, G)$ and $H^m(Q_n, G)$ are zero. This is proved by induction on n. The statement is trivial for Q_0. Suppose the statement has been proved for Q_{n-1}. But Q_n is the cone over Q_{n-1}. By Example VIII 8 the groups $H^m(Q_n \bmod Q_{n-1}, G)$ are zero for all m. By G), $H^m(Q_n, G)$ is isomorphic to $H^m(Q_{n-1}, G)$, and hence zero, and similarly $H_m(Q_n, G)$ is zero.

Example VIII 11. Let R_n be the complex made up of all the faces of dimension $\leqq n$ of an $(n+1)$-simplex. We call this complex the *elementary n-sphere*. If a definite orientation s_{n+1} of the $(n+1)$-simplex is selected we then speak of the *oriented elementary n-sphere* R_n, and we call the oriented faces of s_{n+1} the *positively* oriented simplexes of R_n. (R_n is obviously an orientable pseudo-manifold.) R_n is, of course, a subcomplex of Q_{n+1}, and since, by Proposition D), for $m \leqq n-1$ the groups $H^m(R_n, G)$ and $H_m(R_n, G)$ are respectively isomorphic to $H^m(Q_{n+1}, G)$ and $H_m(Q_{n+1}, G)$, we get from Example VIII 10 that $H^m(R_n, G) = H_m(R_n, G) = 0$, $m \leqq n-1$. Now consider $H^n(R_n, G)$. We could use Example VIII 5 to conclude that $H^n(R_n, G)$ is isomorphic to G, but much simpler is the following argument. The n-cocycles of R_n are just the n-chains of R_n, and if an n-chain φ of R_n is cobounding, the n-chain* $\Phi = h_{Q_{n+1}}\varphi$ must be a coboundary and hence a cocycle of Q_{n+1} (since R_n consists of all the simplexes of Q_{n+1} of dimension $\leqq n$); vice versa every n-cocycle in Q_{n+1} is cobounding (by Example VIII 10). Let s_n^i $(i = 0, 1, 2, \cdots, n+1)$ be the positively oriented n-simplexes of the oriented elementary sphere. If φ is an (n, G)-chain of R_n a necessary and sufficient condition for Φ to be a cocycle in Q_{n+1} is that $\sum_{i=0}^{n+1}\varphi(s_n^i) = 0$, and according to our remark above (that cobounding n-cocycles of R_n correspond to n-cocycles of Q_{n+1}), this equation is also a necessary and sufficient condition for φ to cobound in R_n. It follows that if φ is any (n, G)-chain of R_n, the element $\sum_{i=0}^{n+1}\varphi(s_n^i) (= \sum_{i=0}^{n+1}\Phi(s_n^i))$ of G characterizes the cohomology class of φ and that consequently $H^n(R_n, G)$ is isomorphic to G. The different cohomology classes are represented by the chains $g\, s_n^0$, g being an element of G. Note that to orient R_n means to single out one of the two generators of the free cyclic group $H^n(R_n, \mathfrak{J})$.

Now consider the homology group $H_n(R_n, G)$, which is the same as the group $Z_n(R_n, G)$ of (n, G)-cycles. An n-chain φ is a cycle if and only if $h_{Q_{n+1}}\varphi$ bounds in Q_{n+1}, by the same argument as in the case of co-

* This is the same function φ, but now regarded as a chain of Q_{n+1}: see footnote * on page 115.

homology. Now the elementary $(n + 1)$-chains of Q_{n+1} are the chains $g\, s_{n+1}$; the boundaries of these chains are the functions having the same value g for all simplexes s_n^i and the value $-g$ for all simplexes $-s_n^i$. This proves that $H_n(R_n, G)$ also is isomorphic to G.

Barycentric subdivisions

Let K be a complex and the polytope P its geometric realization. Let P' be the *barycentric subdivision* of P, i.e. the uniquely defined simplicial subdivision of P whose vertices are the vertices of P and in addition the barycenters of the cells of P. Let K' be the vertex-scheme of P'. Then K' is called the *barycentric subdivision* of K. K' can be defined abstractly as follows: we associate a symbol $[s_m]$ with each pair of simplexes s_m and $-s_m$ of K, i.e. with each non-oriented simplex; these symbols $[s_m]$ are the vertices of K'. (The geometric realization of $[s_m]$ is the barycenter of the cell s_m). The non-oriented simplexes of K' are the sequences

$$([s_{m_0}], [s_{m_1}], \cdots, [s_{m_i}])$$

where $m_0 > m_1 > \cdots > m_i$ and s_{m_ν} is a face of $s_{m_{\nu-1}}$.

From the geometric picture P it is clear that to each oriented m-simplex of K belong $(m + 1)!$ oriented m-simplexes of K'. Abstractly the simplex

$$([(p_0, p_1, \cdots, p_m)], [(p_1, \cdots, p_m)], \cdots, [(p_m)]),$$

oriented as indicated, is called an oriented *m-subsimplex* of the oriented simplex $(p_0, \cdots, p_m)$ of K and its negative an oriented subsimplex of $-(p_0, \cdots, p_m)$.

If $s_m' \in K'$ is an oriented subsimplex of the simplex $s_m \in K$ we write $s_m' <\cdot\, s_m$.

Given a complex K, the result of iterating the process of barycentric subdivision is a sequence of complexes called the *successive barycentric subdivisions of* K, and denoted by $K', K'', \cdots, K^{(m)}, \cdots$. We naturally say of an n-simplex $s_n^{(m)} \in K^{(m)}$ that it is an *oriented subsimplex* of $s_n \in K$ if there exists a chain of simplexes $s_n^{(i)} \in K_n^{(i)}$, $i = 1, \cdots, m - 1$, for which $s_n^{(m)} <\cdot\, s_n^{(m-1)} <\cdot \cdots <\cdot\, s_n' <\cdot\, s_n$.

We now prove that the process of barycentric subdivision does not affect homology and cohomology groups.

Let us assign to each chain φ of K' the chain $\Phi = \pi\varphi$ of K defined by

$$\Phi(s_m) = \sum_{s_m' <\cdot s_m} \varphi(s_m').$$

We readily see that the operator π commutes with the coboundary operator: $\gamma\pi\varphi = \pi\gamma\varphi$. Hence (Proposition C)) π induces a homomorphism of $H^m(K', G)$ in $H^m(K, G)$. Moreover

H) *The homomorphism of $H^m(K', G)$ in $H^m(K, G)$ induced by π is an isomorphism of $H^m(K', G)$ on $H^m(K, G)$.*

The argument used in the proof of G) shows that H) is contained in the following Proposition:

I) Let φ be an (m, G)-cocycle of K' and Ψ an $(m - 1, G)$-chain of K such that $\pi\varphi = \gamma\Psi$. Then there exists an $(m - 1, G)$-chain ψ of K' with $\pi\psi = \Psi$ and $\gamma\psi = \varphi$, $m=1, 2, \cdots$.

Proof of I), by induction on the dimension of K. Proposition I) is evident if $\dim K = 0$, for in this case $K' \equiv K$. We shall demonstrate I) for $\dim K = n > 0$ by assuming that I), and consequently H), is valid for any complex of dimension $\leqq n - 1$. We denote by K_{n-1} the subcomplex of K made up of simplexes of dimension $\leqq n - 1$ and by K'_{n-1} the barycentric subdivision of K_{n-1} (K'_{n-1} is, of course, a subcomplex of K'). Our first step will consist in computing the groups

$$H^m(K' \bmod K'_{n-1}) = H^m(K' \bmod K'_{n-1}, G), \quad m = 0, 1, 2, \cdots, n.$$

Let $\pm s_n^1, \pm s_n^2, \cdots, \pm s_n^r$ be all the n-simplexes of K; let Q_n^i, $i = 1, 2, \cdots, r$, be the complex formed by the simplex s_n^i together with its faces, R_{n-1}^i the elementary $(n - 1)$ sphere made up of the faces of s_n^i of dimension $\leqq n - 1$; and $Q_n^{i'}$ and $R_{n-1}^{i'}$ the barycentric subdivisions of Q_n^i and R_{n-1}^i. By Example VIII 9, $H^m(K' \bmod K'_{n-1})$ is the direct sum of the groups

$$H^m(Q_n^{i'} \bmod R_{n-1}^{i'}), \qquad i = 1, 2, \cdots, r. \tag{3}$$

Since $Q_n^{i'}$ is the cone over $R_{n-1}^{i'}$, $H^m(Q_n^{i'} \bmod R_{n-1}^{i'})$ is (Example VIII 8) isomorphic to the group $H^{m-1}(R_{n-1}^{i'})$, which, by the hypothesis of induction, is in turn isomorphic to the group $H^{m-1}(R_{n-1}^i)$ (under the homomorphism established by the operator π). We conclude from Example VIII 11 that $H^m(Q_n^{i'} \bmod R_{n-1}^{i'})$ is zero for $m < n$ and isomorphic to G for $m = n$; moreover, the cohomology class of an n-chain φ of $Q_n^{i'} \bmod R_{n-1}^{i'}$ is characterized by the element $\sum_{s_n' < \cdot s_n^i} \varphi(s_n')$ of G. Observing further that for an n-chain φ of K' the relation $\pi\varphi = 0$ is equivalent to $\sum_{s_n' < \cdot s_n^i} \varphi(s_n') = 0$ for each $i = 1, 2, \cdots, r$, we get the following statement:

J) *The groups* $H^m(K' \bmod K'_{n-1})$ *are zero for* $m < n$. *An* (n, G)-*chain* φ *of* K' *mod* K'_{n-1} *is cohomologous to* 0 *if and only if* $\pi\varphi = 0$.

Now we continue with the demonstration of I). First we consider the case $m = n$. Let ψ_1 be an $(n - 1)$-chain of K' with $\pi\psi_1 = \Psi$ (such a chain can certainly be found, for π is a homomorphism of $C_n(K')$ *on* $C_n(K)$). Then we have

$$\pi(\varphi - \gamma\psi_1) = \pi\varphi - \gamma\pi\psi_1 = \pi\varphi - \gamma\Psi = 0.$$

This means, by the second part of J), that $h_{K'-K'_{n-1}}(\varphi - \gamma\psi_1)$ is cohomologous to 0; in other words,

$$\varphi - \gamma\psi_1 = \gamma\psi_2, \tag{4}$$

where ψ_2 is an $(n - 1)$-chain of K' satisfying $\psi_2(s'_{n-1}) = 0$ if $s'_{n-1} \varepsilon K'_{n-1}$. This implies $\pi\psi_2 = 0$ and together with (4) shows that the chain $\psi = \psi_1 + \psi_2$ satisfies the requirements of I).

Now suppose $m \leqq n - 1$. By applying the hypothesis of induction to the complex K_{n-1} we establish first the existence of an $(m - 1)$-chain ψ_1 of K'_{n-1} such that $\pi\psi_1 = h_{K_{n-1}}\Psi$ and $\gamma\psi_1 = h_{K'_{n-1}}\varphi$. By the first part of J) the hypotheses of F) are satisfied and hence we can construct a chain ψ of K' which is an extension of ψ_1 and has φ for its coboundary. Obviously ψ satisfies the requirements of I). This concludes the proof of I), and therefore of H).

We now consider the operator π' assigning to each (m, G)-chain Φ of K the (m, G)-chain $\pi'\Phi$ of K' defined as follows: $\pi'\Phi(s'_m) = \Phi(s_m)$ if $s'_m <\cdot\, s_m$, and $\pi'\Phi(s'_m) = 0$ if s'_m is not a subsimplex of any s_m. It is clear that π' establishes an isomorphism of $C_m(K)$ in $C_m(K')$. One verifies easily that π' commutes with the boundary operator: $\pi'\beta\Phi = \beta\pi'\Phi$ for any m-chain Φ of K, $m \geqq 0$. Consequently π' leads to a homomorphism of the group $H_m(K, G)$ in the group $H_m(K', G)$. In analogy with H) we have

H′) *The homomorphism of the homology group* $H_m(K, G)$ *induced by the operator* π' *is an isomorphism of the group* $H_m(K, G)$ *on the group* $H_m(K', G)$.

The proof* of H′) is quite analogous to the proof of H). The argument used in the proof of G′) shows that H′) is contained in

I′) *Let* Φ *be an* $(m - 1, G)$-*cycle of* K *and* ψ *an* (m, G)-*chain of* K'

* H′) could also be deduced from H) by application of the duality methods developed below in section 2. See footnote ‡ on page 130.

such that $\beta\psi = \pi'\Phi$. Then there exists an (m, G)-chain Ψ of K such that $\pi'\Psi$ is homologous to ψ, and consequently $\beta\Psi = \Phi$.

PROOF OF I′). I′) is trivial if $\dim K = 0$. Let $\dim K = n > 0$ and assume I′) to hold for any complex of dimension $\leqq n - 1$. Using this hypothesis of induction we establish, in complete analogy with J),

J′) For $m < n$ the groups $H_m(K' \bmod K'_{n-1})$ are zero. An (n, G)-chain φ of K' mod K'_{n-1} is a cycle if and only if $h_{K'}\varphi$ is an image of some chain of K under the homomorphism π'.

We omit the proof of J′). Now we demonstrate I′), first for the case $m = n$. Since $\pi'\Phi(s'_{n-1}) = 0$ if s'_{n-1} is not in K'_{n-1}, the relation $\beta\psi = \pi'\Phi$ implies $h_{K'-K'_{n-1}}\psi$ is a cycle mod K'_{n-1}. By the second part of J′) we conclude that there exists a chain Ψ of K with $\pi'\Psi = \psi$.

Suppose now $m \leqq n - 1$. By F′) and J′) there exists an (m, G)-chain ψ_1 of K'_{n-1} such that $h_{K'}\psi_1$ is homologous to ψ. By the hypothesis of induction applied to the complex K_{n-1} there exists an m-chain Ψ of K_{n-1} such that $\pi'\Psi$ is homologous to ψ_1. This means that $\pi' h_K\Psi$ is homologous to $h_{K'}\psi_1$ and consequently to ψ, thus satisfying the requirement of I′).

REMARK. H) and H′) are special cases of the fundamental theorem of combinatorial topology (Proposition 4 E)): *two complexes having homeomorphic geometric realizations have isomorphic cohomology groups and isomorphic homology groups.* It is this theorem which gives to the cohomology and homology groups, as contrasted with the groups of cycles, cocycles, bounding cycles, etc., their topological meaning.

EXAMPLE VIII 11.1. Let $R_n = R_n^{(0)}, R_n', R_n'', \cdots$ be the elementary n-sphere and its successive barycentric subdivisions. From H) and H′) it follows that the homology and cohomology groups of all these complexes are the same, i.e. (see Example VIII 11) $H^m(R_n^{(i)}, G) = H_m(R_n^{(i)}, G) = 0$ if $m < n$ and $H^n(R_n^{(i)}, G) = H_n(R_n^{(i)}, G) = G$.

An extension of the notion of barycentric subdivision is that of *barycentric subdivision of a complex K modulo a subcomplex L.* We confine ourselves to defining this sort of subdivision geometrically, in terms of the polytopes P and Q which are the geometric realizations of K and L, leaving to the reader the abstract formulation. By a barycentric subdivision of a polytope P mod Q, Q a subpolytope of P, we mean the uniquely determined subdivision of P whose vertices are the vertices of P and in addition the barycenters of those cells of P which are not contained in Q (so that Q is not subdivided). The operators π and π' can be defined in the same way as before and Propositions H) and H′) remain true. It is clear that we can speak of *successive barycentric subdivisions* of K mod L.

2. Duality

Definition VIII 8. Let $\mathfrak{R}_1$ denote the group of real numbers modulo 1. Let G be an arbitrary (Abelian) group. A homomorphism of G in $\mathfrak{R}_1$ is called a *character* of G. If χ_1 and χ_2 are two characters of G, the character χ defined by

$$\chi(g) = \chi_1(g) + \chi_2(g), \qquad g \,\varepsilon\, G,$$

is called the *sum* of χ_1 and χ_2. With this definition of addition the characters of G form a group; this group is called the *character group* of G and denoted by G^*. The zero character of G is the homomorphism sending all of G into the zero of $\mathfrak{R}_1$.

REMARK. If G is a countable discrete group we shall topologize G^* by defining convergence as follows: $\chi_n \to \chi$ in G^* if, for each g in G, the relation $\chi_n(g) \to \chi(g)$ holds in $\mathfrak{R}_1$. It turns out† that the topological space obtained in this way is compact and separable; moreover the group of *continuous* characters of G^* is discrete and isomorphic to G.

Throughout §2 we shall take G as a countable discrete group and its character group topologized as in the Remark immediately above.

EXAMPLE VIII 12. Let $G = \mathfrak{J}$. Then G^* is isomorphic to $\mathfrak{R}_1$.

EXAMPLE VIII 13. Let G be a group of finite order. Then G^* is isomorphic to G.

Definition VIII 9. Let H be a subgroup of G. Those characters of G which send all elements of H into the zero of $\mathfrak{R}_1$, together form a closed subgroup of G^* called the *annihilator* of H. Conversely, given a subgroup J of G^* those elements of G which are sent into the zero of $\mathfrak{R}_1$ by each of the members of J form a subgroup of G called the *annihilator* of J.

Observe that the annihilator of a subgroup of G is a subgroup of G^*, the annihilator of a subgroup of G^* is a subgroup of G, and the smaller the subgroup the larger its annihilator.

The most important property of annihilators is contained in the following Proposition A) due to Pontrjagin,‡ which we state here without proof:

A) a) Let H be a subgroup of G and J a closed subgroup of G^*. Then J is the annihilator of H if and only if H is the annihilator of J. b) In this case $G^* - J$ is the character group of H, and J is the character group of $G - H$.

† Pontrjagin, p. 128, Theorem 31, and p. 134, Theorem 32.

‡ Pontrjagin, p. 135, Theorems 33 and 34.

As a simple corollary of Proposition A) we prove

B) Let H be a subgroup of G, L a subgroup of H, $A(H)$ the annihilator of H and $A(L)$ the annihilator of L. Then $A(L) - A(H)$ is the character group of $H - L$.

Proof. It follows from Proposition A) that $A(L)$ is the character group of $G - L$. Applying Proposition A) again, this time to the group $G - L$ and its subgroup $H - L$, we have the statement.

Note also the following corollary of the first part of Proposition A).

C) Let g be fixed. If $\chi(g) = 0$ for arbitrary χ then $g = 0$, i.e. the annihilator of G^* is the zero of G.

Definition VIII 10. Let us consider two groups G_1 and G_2 and a homomorphism h of G_1 in G_2. To each character χ of G_2 we assign a character of G_1, denoted by $h^*\chi$, according to the formula

$$\{h^*\chi\}(g_1) = \chi(h(g_1)),$$

g_1 denoting the generic element of G_1. By this means we obtain a homomorphism h^* of G_2^* in G_1^*, and this homomorphism h^* is said to be *dual* to h.

D) Let G_1 and G_2 be two groups and h a homomorphism of G_1 in G_2. Then the annihilator of the subgroup $h(G_1)$ of G_2 is the kernel of the homomorphism h^* dual to h. Further the annihilator of the group $h^*(G_2^*)$ is the kernel of h.

Proof. The first part follows immediately from Definition VIII 10, since $\chi(h(g_1)) = 0$ means $\{h^*\chi\}(g_1) = 0$. To prove the second part observe that the annihilator of $h^*(G_2^*)$ consists of those elements $g_1 \varepsilon G_1$ for which

$$\{h^*\chi\}(g_1) = \chi(h(g_1)) = 0 \text{ for arbitrary } \chi \text{ in } G_2^*;$$

and this means, by C), that $h(g_1) = 0$.

E) Not only does h determine h^*, but also h^* determines h; i.e. the correspondence between a homomorphism and its dual is 1:1.

Proof. For let h and h' be two homomorphisms of G_1 in G_2 and h^* and h'^* their duals. Consider the homomorphism $h - h'$ of G_1 in G_2, defined in the natural way by

$$\{h - h'\}(g_1) = h(g_1) - h'(g_1).$$

It is easy to see that the dual of $h - h'$ is $h^* - h'^*$. Now suppose

$h^* = h'^*$. Then (see Definition VIII 10) for each character χ of G_2 and $g_1 \varepsilon G_1$,

$$0 = \chi(\{h - h'\}(g_1)).$$

By C), therefore $\{h - h'\}(g_1) = 0$, i.e. $h = h'$. Consequently it is impossible for different h and h' to have the same dual.

F) In order that h be a homomorphism of G_1 *on* G_2 it is necessary and sufficient that the homomorphism h^* of G_2^* in G_1^* be an *iso*morphism.

PROOF. Recalling that a homomorphism is an isomorphism if and only if its kernel is zero, we see from the first part of D) that h^* is an isomorphism if and only if the annihilator of $h(G_1)$ is zero. But by A) a) this is equivalent to the statement that $h(G_1)$ is the annihilator of the zero of G_2^*, i.e. $h(G_1) = G_2$.

DUALITY BETWEEN HOMOLOGY AND COHOMOLOGY GROUPS OF A COMPLEX

We shall now employ the technique of group duality in the study of homology and cohomology properties of complexes. Let K be a complex, G a countable group, and $n \geqq -1$. Let C_n be the group of (n, G)-chains of K. Let f be a character of C_n. For a fixed n-simplex s_n^0 of K, f applied to the elementary chains gs_n^0 gives a character of G. Denote this character by $\chi = \chi(s_n^0)$. Then evidently

$$\chi(-s_n^0) = -\chi(s_n^0).$$

Hence χ is an (n, G^*)-chain (Definition VIII 2). The reader will easily verify that in this way an isomorphism is established between the group $C_n(K, G^*)$ and the character group of $C_n(K, G)$. If φ is an (n, G)-chain and ψ an (n, G^*)-chain the element $\psi(\varphi)$ of $\mathfrak{R}_1$ assigned to φ by the character ψ can be computed as follows: Let $\pm s_n^i$, $i = 1, \cdots, k$, be the oriented n-simplexes of K. Then

$$\psi(\varphi) = \sum_{i=1}^{k} \{\psi(s_n^i)\}(\varphi(s_n^i)). \tag{1}$$

We now assert that if φ^* is an $(n + 1, G^*)$-chain and φ an (n, G)-chain, then

$$\{\beta\varphi^*\}(\varphi) = \{\varphi^*\}(\gamma\varphi). \tag{2}$$

The reader will verify (2) by applying (1). Formula (2) shows that the homomorphism β of $C_{n+1}(K, G^*)$ in $C_n(K, G^*)$ is dual (Definition VIII 10) to the homomorphism γ of $C_n(K, G)$ in $C_{n+1}(K, G)$. Now we make use of Proposition D), noting that the kernel of β is the group $Z_{n+1}(K,G^*)$ of $(n + 1, G^*)$ cycles, the kernel of γ is the group $Z^n(K, G)$ of (n, G)-cocycles, the image of $C_{n+1}(K, G^*)$ under β is the group $B_n(K, G^*)$ of bounding (n, G^*)-cycles, and the image of $C_n(K, G)$ under γ is the group $B^{n+1}(K, G)$ of cobounding $(n + 1, G)$-cocycles. We obtain the result (lowering dimension by one in part a)):

G) a) $Z_n(K, G^*)$ is the annihilator of $B^n(K, G)$,
b) $Z^n(K, G)$ is the annihilator of $B_n(K, G^*)$.

Using the first part of Proposition A) we can replace b) by
b′) $B_n(K, G^*)$ is the annihilator of $Z^n(K, G)$.

A consequence is the following result, which was the main goal of this section:

H) $H_n(K, G^*)$ *is the character-group of* $H^n(K, G)$. *In particular* $H_n(K, \mathfrak{R}_1)$ *is the character group of* $H^n(K, \mathfrak{I})$.

Proof. Apply Propositions B) and G), taking $C_n(K, G)$ for G, $Z^n(K, G)$ for H, and $B^n(K, G)$ for L.

Remark 1. Using similar methods we could prove that $H^n(K, G^*)$ is the character-group of $H_n(K, G)$.

Remark 2. $H_n(K, G^*)$ and $H^n(K, G^*)$, as the character groups of the countable groups $H^n(K, G)$ and $H_n(K, G)$, have compact separable topologies (see Remark on page 124). As a matter of fact these topologies can be defined directly, using the topology of G^* and without referring to the groups $H^n(K, G)$ and $H_n(K, G)$.

Remark 3. In the connectivity theory of finite complexes no special preference can be given to cohomology over homology. The situation is different, however, for the connectivity theory of more general spaces, where there is a distinct advantage in the use of cohomology.

3. Simplicial mappings of complexes

Definition VIII 11. Let K_1 and K_2 be two complexes. A mapping f of the vertices of K_1 in those of K_2 is *simplicial* if for any n-simplex of K_1 with vertices $p_0, \cdots, p_n$ the points $f(p_0), \cdots, f(p_n)$ are vertices of a simplex of K_2 (of dimension possibly lower than n). Let $s_n = (p_0, \cdots, p_n)$ be an oriented n-simplex of K_1 and f a simplicial mapping of K_1 in K_2. If all the $f(p_0), \cdots, f(p_n)$ are distinct we denote by $f(s_n)$ the oriented n-simplex $(f(p_0), \cdots, f(p_n))$; $f(s_n)$ is not defined otherwise.

A) Let f be a simplicial mapping of K_1 in K_2. Then† f induces a homomorphism h^f of $H^n(K_2)$ in $H^n(K_1)$ as follows: Given an n-chain ψ of K_2 we associate with it the n-chain $\varphi = h^f\psi$ of K_1 by setting

$$\begin{cases} \varphi(s_n) = \psi(f(s_n)) \text{ if } f(s_n) \text{ is defined,} \\ \varphi(s_n) = 0 \text{ otherwise.} \end{cases}$$

h^f is plainly a homomorphism of $C_n(K_2)$ in $C_n(K_1)$. It is easy to verify that h^f commutes with γ. Hence (Proposition 1C)), h^f leads to a homomorphism, which we continue to denote by the same symbol h^f, of $H^n(K_2)$ in $H^n(K_1)$.

B) Similarly, a simplicial mapping f of K_1 in K_2 induces a homomorphism h_f of $H_n(K_1)$ in $H_n(K_2)$ as follows: Let z_n be any n-simplex of K_2. Then, given any n-chain φ of K_1 we associate with it the n-chain $\psi = h_f\varphi$ of K_2 by setting

$$\psi(z_n) = \sum_{f(s_n)=z_n} \varphi(s_n).$$

h_f is plainly a homomorphism of $C_n(K_1)$ in $C_n(K_2)$. It is easy to see that h_f commutes with β. Hence (Proposition 1 C)), h_f leads to a homomorphism, which we continue to denote by h_f, of $H_n(K_1)$ in $H_n(K_2)$.

The reader will easily prove that if f maps K_1 simplicially in K_2 and g maps K_2 simplicially in K_3, then the homomorphism h_{gf} of $H_n(K_1)$ in $H_n(K_3)$ induced by gf is the product h_gh_f. The analogous statement of transitivity holds for the homomorphisms of cohomology groups induced by simplicial mappings.

C) Let G^* be the character group of G, f a simplicial mapping of K_1 in K_2, h_f the homomorphism of $C_n(K_1, G^*)$ in $C_n(K_2, G^*)$ induced by f, and h^f the homomorphism of $C_n(K_2, G)$ in $C_n(K_1, G)$ induced by f. Now $C_n(K_1, G^*)$ and $C_n(K_2, G^*)$ are the character groups of $C_n(K_1, G)$ and $C_n(K_2, G)$. We assert that h^f and h_f are dual (Definition VIII 10), i.e. if ψ is an (n, G^*)-chain of K_1 and φ an (n, G)-chain of K_2,

$$\{h_f\psi\}(\varphi) = \psi(h^f(\varphi)). \tag{1}$$

We leave the easy proof to the reader.

D) Let G be a countable group and G^* its character group. We recall that $H_n(K_1, G^*)$ and $H_n(K_2, G^*)$ are the character groups of $H^n(K_1, G)$ and $H^n(K_2, G)$. If in (1) we take ψ as a cycle and φ as a

† The basis group G is arbitrary.

cocycle we see that the homomorphism h_f of $H_n(K_1, G^*)$ in $H_n(K_2, G^*)$ is dual to the homomorphism h^f of $H^n(K_2, G)$ in $H^n(K_1, G)$. Using similar methods one can prove that f induces dual homomorphisms of $H^n(K_2, G^*)$ in $H^n(K_1, G^*)$ and $H_n(K_1, G)$ in $H_n(K_2, G)$.

Example VIII 14. Let L be a subcomplex of K and f the identity mapping of L in K. Then f is simplicial. The homomorphisms of the cohomology groups of K in those of L, and of the homology groups of L in those of K are, of course, the natural homomorphisms of these groups (Definition VIII 7).

Example VIII 15. Let K be a complex and K' its barycentric subdivision. On page 120 we defined a homomorphism π of the n-chains of K' in the n-chains of K, and we know (Proposition 1 H)) that this homomorphism is an isomorphism of $H^n(K', G)$ on $H^n(K, G)$.

Let us now assign to each vertex p of K' a definite, although arbitrarily chosen, vertex of the simplex of K of which p is the barycenter. This gives us a simplicial mapping f of K' in K, and consequently a homomorphism h^f (see A)) of the (n, G)-chains of K in the (n, G)-chains of K'. One easily sees that πh^f acting on $C_n(K, G)$ is the identity transformation. Hence πh^f acting on $H^n(K, G)$ is also the identity transformation, and since, as remarked above, π is an isomorphism of $H^n(K', G)$ on $H^n(K, G)$, it must likewise be true that h^f is an isomorphism of $H^n(K, G)$ on $H^n(K', G)$. Thus we have the result that the homomorphism of $H^n(K, G)$ in $H^n(K', G)$ induced by f is actually the isomorphism of $H^n(K, G)$ on $H^n(K', G)$ inverse to π.

Similarly we can prove that the homomorphism of $H_n(K', G)$ in $H_n(K, G)$ induced by f is actually the isomorphism of $H_n(K', G)$ on $H_n(K, G)$ inverse to π'. Moreover if G is a countable group then π' acting on $H_n(K, G^*)$ is the dual homomorphism to π acting on $H^n(K', G)$

Example VIII 16. Let K be a complex and R an elementary n-sphere (see Example VIII 11) or one of its successive barycentric subdivisions. Let f be a simplicial mapping of K in R and $\mathfrak{J}$ the group of the integers. Then we know that $H^n(R, \mathfrak{J})$ is a free cyclic group, the generator being determined by the orientation of the sphere. Hence the homomorphism h^f of $H^n(R, \mathfrak{J})$ in $H^n(K, \mathfrak{J})$ is completely determined by the element $h^f(\alpha)$ corresponding to the generator α of $H^n(R, \mathfrak{J})$; this element is called the *degree* of f. The degree of f is easily computed. Let z_n^0 be an arbitrary positively oriented n-simplex† of R. By Example VIII 11 the elementary chain $1z_n^0$ is a basis

† I.e. an oriented subsimplex of one of the positively oriented simplexes of the elementary n-sphere.

for the group of n-cocycles of R. Then the degree of f is the cohomology class of the following chain φ of K: For each n-simplex s_n of K,

$$\begin{cases} \varphi(s_n) = 1 & \text{if } f(s_n) = z_n^0, \\ \varphi(s_n) = -1 & \text{if } f(s_n) = -z_n^0, \\ \varphi(s_n) = 0 & \text{otherwise.} \end{cases}$$

If K also is an elementary n-sphere or one of its barycentric subdivisions, the degree of f may be regarded as an integer, namely $\sum\varphi(s_n)$, the sum being taken over all positively-oriented s_n; intuitively the degree is the algebraic number of times z_n^0 is covered, and for each integer m it is easy to construct an f having m as degree.

Let K be arbitrary again and consider the character-group $\mathfrak{R}_1$ of $\mathfrak{J}$. The simplicial mapping f induces a homomorphism h_f of $H_n(K, \mathfrak{R}_1)$ in $H_n(R, \mathfrak{R}_1)$. We know that $H_n(R, \mathfrak{R}_1)$ is isomorphic to $\mathfrak{R}_1$, so that h_f is a character of $H_n(K, \mathfrak{R}_1)$. But h_f and h^f are dual and hence determine each other. Hence we could just as well specify the degree of f by this character† of the group $H_n(K, \mathfrak{R}_1)$.

REMARK. A degree can be defined in precisely the same way for the more general case of a simplicial mapping of an n-complex in an arbitrary n-dimensional orientable pseudo-manifold (see Example VIII 5). If K itself is an orientable pseudo-manifold then the degree can be regarded as an integer.

E) Suppose f and g are two simplicial mappings of K_1 in K_2 with this property: for each simplex s of K_1, $f(s)$ and $g(s)$ are faces of a common simplex of K_2. Then f and g yield the same homomorphism of the homology groups of K_1 in those of K_2 and of the cohomology groups of K_2 in those of K_1.

PROOF. It is sufficient to demonstrate E) for cohomology groups, for given any countable group G the homomorphism induced by f of the cohomology group $H^n(K_2, G^*)$ in $H^n(K_1, G^*)$ is dual (Proposition D)) to the homomorphism induced by f of $H_n(K_1, G)$ in $H_n(K_2, G)$, and hence (Proposition 2 E)) the second is determined by the first. The case of an arbitrary basis group G can be reduced to the case of a countable basis group‡ by the following simple observation: Let φ be an (n, G)-cycle of K and let G_1 be the countable subgroup of G gener-

† This is also in accordance with the fact that $H^n(K, \mathfrak{J})$ is the group of continuous characters of its character group $H_n(K, \mathfrak{R}_1)$ (see the Remark on page 124).

‡ A similar argument could have been used to derive Proposition 1 H') from Proposition 1 H).

ated by the elements $\varphi(s_n)$. Then $h_f\varphi$ and $h_g\varphi$ can be regarded as (n, G_1)-cycles, and if they are homologous with respect to the basis group G_1 they are, a fortiori, homologous with respect to the basis group G.

We may assume that f and g differ on only one vertex of K_1, since g can be gotten from f by a finite sequence of modifications each consisting in changing the image of only one vertex. In accordance with this, let p be the only vertex for which $f(p)$ differs from $g(p)$. We associate with each (n, G)-chain φ of K_2 an $(n - 1, G)$-chain φ^* in K_1 as follows: Suppose s_{n-1} is an $(n - 1)$-simplex of K_1 having p as a vertex and such that the simplex $f(s_{n-1})$ is defined and does not have $g(p)$ as a vertex. Let z_n be the n-simplex in K_2 whose vertices are $g(p)$ followed by the vertices of $f(s_{n-1})$. Then we set $\varphi^*(s_{n-1}) = \varphi(z_n)$. For all other $(n - 1)$-simplexes of K_1 we set $\varphi^*(s_{n-1}) = 0$. Denote by h^f the homomorphism induced by f of the (n, G)-chains of K_2 in the (n, G)-chains of K_1. Then a simple computation shows that

$$\gamma(\varphi^*) = -(\gamma\varphi)^* + h^f\varphi - h^g\varphi.$$

If φ is a cocycle the first term on the right is zero, so that this formula establishes a cohomology between $h^f\varphi$ and $h^g\varphi$, and proves the Proposition.

4. Homology and cohomology in compact spaces

One of the great achievements of topology in recent years has been the extension of homology theory from polytopes to arbitrary compact spaces.† This was one of the most essential steps in bridging the gap between the combinatorial and point-set methods in topology.

Direct and inverse systems of groups

Let Σ be a *partially ordered set*, i.e. a set in which we have a transitive

† Our presentation is closely related to the formulation of the Čech homology theory of Steenrod's paper: Universal Homology Groups, *American Journal of Mathematics* 58 (1936), pp. 661–701, and also to Freudenthal's exposition: Alexanderscher und Gordonscher Ring und ihre Isomorphie, *Annals of Mathematics* 38 (1937), pp. 647–655. Čech's work: Théorie générale de l'homologie dans un espace quelconque, *Fundamenta Mathematicae* 19 (1932), pp. 149–184, in turn is based on Alexandroff's ideas of spectrum and nerve of a covering: Untersuchungen über Gestalt und Lage abgeschlossener Mengen beliebiger Dimension, *Annals of Mathematics* 30 (1929), pp. 71–85. In contrast with the procedures of Alexandroff and Čech, the older Vietoris theory uses infinite complexes whose vertices are the points of the space.

See also the recent paper of Alexandroff (which appeared after this book was in press): General combinatorial topology, *Trans. Am. Math. Soc.* 49 (1941), 41–105.

relation $<$ defined* for some (but not necessarily all) pairs of elements. A subset Σ' of Σ is said to be *confinal* with Σ if every element of Σ has a successor in Σ': given an element σ of Σ there is an element σ' of Σ' such that $\sigma < \sigma'$. Σ is called a *directed set* if every pair of elements have a common successor: given σ and τ in Σ there is a ρ in Σ satisfying $\sigma < \rho$ and $\tau < \rho$.

Definition VIII 12. Let Σ be a directed set. Suppose to each element σ of Σ is assigned a group G_σ and to each pair of elements $\sigma < \tau$ of Σ there corresponds a homomorphism $h^{\sigma\tau}$ of G_σ in G_τ such that if $\rho < \sigma < \tau$ then

$$h^{\rho\tau} = h^{\sigma\tau}(h^{\rho\sigma}).$$

A system of groups of this sort is called a *direct system of groups*. Given a direct system of groups we now define a new group called the *limit group G of the direct system*: An element g_σ of a group G_σ is said to be *related* to an element g_τ of G_τ if they have a common successor: i.e. if there is a ρ, satisfying $\sigma, \tau < \rho$, such that $h^{\sigma\rho}g_\sigma = h^{\tau\rho}g_\tau$. A class consisting of all the elements related to an element g_σ of G_σ is defined to be an *element* of G, and g_σ is called a *representative*† in G_σ of this element of the limit group. *Addition* in G is defined as follows: Let g_σ and g_τ be representatives of two elements g and g' of G and ρ an element of Σ for which $\sigma, \tau < \rho$. Then by the *sum* of g and g' we mean the element of G determined by the element $h^{\sigma\rho}g_\sigma + h^{\tau\rho}g_\tau$ of G_ρ. (The zero of G is the class containing the zero of each one of the groups G_σ.)

EXAMPLE VIII 17. Let $G_1, G_2, \cdots$ be a sequence of groups where G_n is a subgroup of G_{n+1}, and let h^{mn}, $m \leqq n$, denote the identity mapping of G_m in G_n. Then $\{G_n\}$ is obviously a direct system of groups and the elements of the limit group make up the set-theoretical sum of the G_n.

EXAMPLE VIII 18. Let $\{G_\sigma\}$ be a direct system of groups such that $h^{\sigma\tau}$ is an isomorphism of G_σ on G_τ. Then the limit group of $\{G_\sigma\}$ is isomorphic to each G_σ.

A) Suppose Σ is a directed set, Σ' a confinal subset and $\{G_\sigma\}$, $\sigma \in \Sigma$, a direct system of groups. Then the subsystem obtained by restricting σ to Σ' is obviously a direct system, and it is readily seen that its limit group is isomorphic to the limit group of the entire system.

B) If in the direct system $\{G_\sigma\}$, $\sigma \in \Sigma$, each G_σ is countable and if Σ

* We do not assume that $\sigma < \tau$ and $\tau < \sigma$ imply σ identical to τ.

† An element of G need not have representatives in each G_σ and may have more than one representative in some G_σ, but an element of G is completely determined by any one of its representatives.

has a countable confinal subset, then the limit group of the direct system is countable.

Definition VIII 13. Let Σ be a directed set. Suppose to each $\sigma \varepsilon \Sigma$ there is assigned a group G_σ and to each pair of elements $\sigma < \tau$ of Σ there corresponds a homomorphism $h_{\tau\sigma}$ of G_τ in* G_σ such that if $\rho < \sigma < \tau$ then

$$h_{\tau\rho} = h_{\sigma\rho}(h_{\tau\sigma}).$$

Such a system is called an *inverse system of groups.* We now define the *limit group G of an inverse system of groups*:† an *element* of G is a collection, consisting of exactly one member g_σ from each G_σ, with the property $h_{\tau\sigma}g_\tau = g_\sigma$ if $\sigma < \tau$; g_σ is called the *representative*‡ of g in G_σ. *Addition* in G is defined as follows: let g and g' be two elements of G and g_σ and g_σ' their representatives in G_σ. Then by the *sum* of g and g' we mean the class of elements $\{g_\sigma + g_\sigma'\}$. (The zero of the limit group is the class consisting of the zero of each of the groups G_σ.)

Example VIII 19. Let $G_1, G_2, \cdots$ be a sequence of groups with G_{n+1} a subgroup of G_n, and let h_{nm}, $m \leqq n$, be the identity homomorphism of G_n in G_m. Then the sequence $\{G_n\}$ is obviously an inverse system of groups and the elements of the limit group make up the set-theoretical intersection of the G_n.

Example VIII 20. Let $\{G_\sigma\}$ be an inverse system of groups such that $h_{\tau\sigma}$ is an isomorphism of G_τ on G_σ. Then the limit group of $\{G_\sigma\}$ is isomorphic to each G_σ.

C) Let Σ be a directed set and Σ' a confinal subset. Suppose $\{G_\sigma\}$, $\sigma \varepsilon \Sigma$, is an inverse system of groups. Then the subsystem obtained by restricting σ to Σ' is obviously an inverse system, and it is readily seen that its limit group is isomorphic to the limit group of the entire system.§

D) Let $\{G_\sigma\}$ be a direct system of groups with homomorphisms $h^{\sigma\tau}$,

* Instead of as before of G_σ in G_τ.

† If G_σ are topological groups forming an inverse system with continuous homomorphisms then the limit group of $\{G_\sigma\}$ can be regarded as a topological group; but if $\{G_\sigma\}$ is a direct system there is no way to derive a topology for the limit group from the topology of the G_σ.

‡ Thus an element of G has exactly one representative in each G_σ, but different elements of G may have the same representative in a particular G_σ.

§ It would be wrong to conclude from C) that, in analogy with B), if each G_σ is countable and Σ has a countable confinal subset, then the limit group is countable.

$\sigma < \tau$. Let G_σ^* be the character group of G_σ and $h_{\tau\sigma}{}^*$ the homomorphism of G_τ^* in G_σ^* dual to $h^{\sigma\tau}$ (see Definition VIII 10). Then a) the groups $\{G_\sigma^*\}$ with homomorphisms $h_{\sigma\tau}{}^*$, $\sigma < \tau$, form an inverse system of groups, and b) if G is the limit group of $\{G_\sigma\}$ its character group G^* is the limit group of $\{G_\sigma^*\}$.

PROOF. We leave the proof of a) to the reader and proceed to the proof of b). By the definition of the limit group of a direct system each element g_σ of G_σ determines an element of G, and this correspondence is plainly a homomorphism, h^σ, of G_σ in G. Further if $\sigma < \tau$

$$h^\sigma = h^\tau h^{\sigma\tau}. \tag{1}$$

Now let χ be a character of G. For a fixed σ, χh^σ is a character of G_σ; i.e. an element, g_σ^*, of G_σ^*. Now from (1) it follows that if $\sigma < \tau$

$$\overset{*}{h}_{\tau\sigma}\overset{*}{g}_\tau = \overset{*}{g}_\sigma ; \tag{2}$$

and hence the class $\{g_\sigma^*\}$ is an element of the limit group of the inverse system $\{G_\sigma^*\}$. This shows that to each character of G corresponds an element of the limit group of $\{G_\sigma^*\}$. The reader will easily complete the proof by showing this correspondence to be an isomorphism of G^* on the limit group of $\{G_\sigma^*\}$.

EXAMPLE VIII 20.1. Let each of the groups G_i, $i = 1, 2, \cdots$, be the group $\mathfrak{J}$ of integers, and let h^{ik}, $i \leqq k$, be the homomorphism of G_i in G_k defined by

$$h^{ik}(m) = m \cdot 2^{k-i}, \qquad m \in \mathfrak{J}. \tag{3}$$

Then the groups $\{G_i\}$ with these homomorphsims form a direct system whose limit group G is easily seen to be isomorphic to the group of dydaically rational numbers, i.e. fractions of the form $m/2^n$. By D) the character group G^* of G is the limit group of the inverse system $\{G_i^*\}$, where G_i^* is isomorphic to $\mathfrak{R}_1$ and the homomorphisms h_{ki}^*, $i \leqq k$, are defined by

$$h_{ki}^*(r) = r \cdot 2^{k-i}, \qquad r \in \mathfrak{R}_1. \tag{4}$$

G^* is called (van Dantzig) the dyadic solenoid; as a character group of a countable group G, it is a compact space and its topological structure is that of an indecomposable one-dimensional continuum.

If we took for G_i the group of integers reduced modulo 2^i with homomorphisms again defined by (3), we would get for G the group of dyadic rational numbers reduced modulo 1 and for its character group G^* the dyadic 0-dimensional group.†

Homology and cohomology in compact spaces

We now consider an arbitrary compact space X. Let Σ be the set of all coverings of X, partially-ordered by the statement: σ precedes τ, $\sigma < \tau$, if τ is a refinement‡ of σ. Σ is a directed set since any two coverings σ and τ have a common refinement (obtained by the mutual intersections of the elements of σ with those of τ). With each covering σ we consider its nerve $N(\sigma)$ (see Definition V 8). Suppose $\sigma < \tau$, i.e. τ is a refinement of σ. Let us select for each member of τ a member of σ containing it. This gives a simplicial mapping of $N(\tau)$ in $N(\sigma)$ which is called a *projection* of $N(\tau)$ in $N(\sigma)$. In general there will be many projections of $N(\tau)$ in $N(\sigma)$, corresponding to the different choices of the element of σ containing each given element of τ; but any two projections p and p' satisfy the condition of Proposition 3 E) and consequently induce the same homomorphism $h^{\sigma\tau}$ of $H^n(N(\sigma), G)$ in $H^n(N(\tau), G)$ and the same homomorphism $h_{\tau\sigma}$ of $H_n(N(\tau), G)$ in $H_n(N(\sigma), G)$. If $\rho < \sigma < \tau$ and p is a projection of $N(\tau)$ in $N(\sigma)$ and q a projection of $N(\sigma)$ in $N(\rho)$ then qp is a projection of $N(\tau)$ in $N(\rho)$; hence $h^{\rho\tau} = h^{\sigma\tau}h^{\rho\sigma}$ and $h_{\tau\rho} = h_{\sigma\rho}h_{\tau\sigma}$. This shows that for any integer n and basis group G, $\{H^n(N(\sigma), G)\}$ is a direct system with homomorphisms $h^{\sigma\tau}$ and $\{H_n(N(\sigma), G)\}$ is an inverse system with homomorphisms $h_{\tau\sigma}$.

Definition VIII 14. Let X be a compact space, G a group, n an integer $\geqq 0$, and $\{\sigma\}$ the collection of all coverings of X. By the (n, G)-*homology group* $H_n(X, G)$ of X (written $H_n(X)$ when no confusion is possible) we mean the limit group of the inverse system $\{H_n(N(\sigma), G)\}$ and by the (n, G)-*cohomology group* $H^n(X, G)$ of X (written $H^n(X)$ when no confusion is possible) we mean the limit group of the direct system $\{H^n(N(\sigma), G)\}$.

Remark 1. It is clear that Definition VIII 14 is topological in character, i.e. homeomorphic spaces have the same homology and cohomology groups.

Remark 2. In the study of complexes we observed complete symmetry between homology and cohomology groups. But here, in the connectivity theory of a compact space, this symmetry disappears, in that homology groups are obtained as limits of *inverse* systems, while cohomology groups are limits of *direct* systems.

† See Pontrjagin: The theory of topological commutative groups, *Ann. Math.* 35 (1934), pp. 361–388.

‡ *Not* σ is a refinement of τ.

Topological invariance of combinatorial homology and cohomology groups

E) Let X be a polytope and K its vertex scheme, $H^n(X)$ and $H_n(X)$ the (topologically invariant) cohomology and homology groups of the space X, and $H^n(K)$ and $H_n(K)$ the (combinatorial) cohomology and homology groups of the complex K. Then *$H^n(X)$ is isomorphic to $H^n(K)$ and $H_n(X)$ is isomorphic to $H_n(K)$. Hence if K_1 and K_2 are vertex schemes of homeomorphic polytopes they have isomorphic cohomology groups and isomorphic homology groups.*

Proof. Let $K = K^{(0)}$, and denote by $K^{(1)}, K^{(2)}, \cdots$ the successive barycentric subdivisions of K. To each $K^{(i)}$ corresponds a simplicial decomposition X^i of X. Let σ_i be the covering of X consisting of the stars (see Definition V 9) of the vertices of X^i. Note† that $N(\sigma_i)$ is $K^{(i)}$. It is clear (V 4C)) that the sequence $\{\sigma_i\}$ is confinal with the set of all coverings, since the mesh of σ_i tends to zero. Hence by A) and C) the groups $H^n(X)$ and $H_n(X)$ are the limit groups of the direct and inverse systems $\{H^n(N(\sigma_i))\}$ and $\{H_n(N(\sigma_i))\}$. But a projection f of $N(\sigma_{i+1})$ in $N(\sigma_i)$ is a simplicial mapping of $K^{(i+1)}$ in $K^{(i)}$ of the type considered in Example VIII 15, and hence yields isomorphisms of $H^n(N(\sigma_i))$ on $H^n(N(\sigma_{i+1}))$ and of $H_n(N(\sigma_{i+1}))$ on $H_n(N(\sigma_i))$. It now follows from Examples VIII 18 and VIII 20 that the limit groups, i.e. $H^n(X)$ and $H_n(X)$, are isomorphic to the groups $H^n(K)$ and $H_n(K)$.

Remark. Let φ be a cocycle of a complex K and φ^* a cocycle of one of the successive barycentric subdivisions $K^{(m)}$ of K (or more generally of one of the successive barycentric subdivisions of K modulo a subpolytope) and suppose that φ and φ^* are related as follows: if s_n is an oriented simplex of K then

$$\varphi(s_n) = \sum \varphi^*(s_n^{(m)}), \tag{3}$$

the sum being extended over all oriented subsimplexes $s_n^{(m)}$ of s_n. This means that φ^* is obtained from φ by iterated application of the operator‡ π. Then from the preceding discussion it follows that φ and φ^* represent the same element of $H^n(P)$, where P denotes the geometrical realization of K.

Example VIII 21. Let I_n be an n-cube. Since I_n is homeomorphic to an n-cell, $H^m(I_n, G) = H_m(I_n, G) = 0$ for all m by Example VIII 10.

† See footnote † on page 69.
‡ Defined on page 120.

Example VIII 22. Let S_n be an n-sphere. Then since S_n is homeomorphic to the polytope made up of the proper faces of an $(n + 1)$-cell, it follows readily from Example VIII 11 that if $m < n$ both $H^m(S_n, G)$ and $H_m(S_n, G)$ are zero, while both $H^n(S_n, G)$ and $H_n(S_n, G)$ are isomorphic to G.

Example VIII 22.1. Let the compact space X be the sum of the closed disjoint sets X_1 and X_2. If $n > 0$ then $H^n(X)$ is the direct sum of the groups $H^n(X_1)$ and $H^n(X_2)$, and $H_n(X)$ is the direct sum of the groups $H_n(X_1)$ and $H_n(X_2)$. This follows easily from Example VIII 6.

F) Let X be a compact space. Then if $n > \dim X$, $H_n(X) = H^n(X) = 0$.

Proof. By Theorem V 1 the set of coverings of X of order $\leqq n$ make up a confinal subset of all the coverings of X. Combining this with Propositions A) and C) and Example VIII 2 proves F).

Duality between homology and cohomology groups of a compact space

Now let G be a countable group. Observe that the set $\{\sigma\}$ of all coverings of X contains a countable confinal subset, for if σ_i is a covering of mesh $1/i$ every covering has one of the σ_i as a refinement. From Proposition B) we see that the cohomology group $H^n(X, G)$ is countable.

G) *Let G be a countable group and G^* its character group. Then $H_n(X, G^*)$ is the character group of $H^n(X, G)$.*

Proof. We have already seen (2H)) that for each covering σ of X, $H_n(N(\sigma), G^*)$ is the character group of $H^n(N(\sigma), G)$. Combining this with Proposition D) the result follows.

An important consequence of Proposition G) and the fact that $H^n(X, G)$ is countable is that we can apply Pontrjagin's duality theory, expounded in §2, to the pair of groups $H^n(X, G)$ and $H_n(X, G^*)$.

Remark 1. Let X be an arbitrary compact space and G a countable group. As the character group of a countable discrete group, $H_n(X, G^*)$ admits a compact topology, but there is no natural way to define a topology in the group $H^n(X, G^*)$.

Remark 2. Let X be a compact space and G an *arbitrary* group. It is known* that the homology and cohomology groups of X with basis G are completely determined by either the homology groups of X with basis $\mathfrak{R}_1$ or dually, by the cohomology groups of X with basis $\mathfrak{J}$.

Remark 3. Formally the definitions of homology and cohomology given above can be applied to non-compact spaces, but the homology theory obtained in this way would be very unsatisfactory. For one, the 1-dimensional homology group of the straight line, even with the simplest groups G, is extremely complicated.† Moreover there is no simple duality between homology and cohomology.

Example VIII 22.2. We define a compact space X as follows: The points of X are sequences of complex numbers (the number ∞ included):

$$(z_1, z_2, \cdots, z_i, \cdots)$$

satisfying

$$z_{i+1}^2 = z_i, \qquad i = 1, 2, \cdots .$$

Convergence in X means term by term convergence. X is clearly a compact separable space. Let i be a fixed integer > 0. By assigning to each point $z = (z_1, z_2, \cdots)$ its ith "coordinate" z_i we obtain a mapping f_i of X in the complex number sphere S. With each covering σ of S we associate a covering σ^i of X whose members are the inverse images of the members of σ under f_i. The nerve $N(\sigma^i)$ is obviously identical with $N(\sigma)$.

One can easily construct a sequence $\{\sigma_i\}$ of coverings of S with mesh converging to 0 such that σ_{i+1}^{i+1} is a refinement of σ_i^i, $i = 1, 2, \cdots$. The reader will observe that the projection mapping of $N(\sigma_{i+1}^{i+1}) = N(\sigma_{i+1})$ in $N(\sigma_i^i) = N(\sigma_i)$ is a simplicial mapping of degree 2, and consequently the projection homomorphisms of $H^2(N(\sigma_i^i), \mathfrak{J})$ in $H^2(N(\sigma_k^k), \mathfrak{J})$ $(k \geqq i)$ are the homomorphisms (3) of Example VIII 20.1. The group $H^2(X, \mathfrak{J})$ is the limit of the direct system $H^2(N(\sigma_i^i), \mathfrak{J})$ and hence by Example VIII 20.1 isomorphic to the group of dyadic rational numbers. Its character group $H_2(X, \mathfrak{R}_1)$ is the dyadic solenoid.

Let Y be the closed subset of X consisting of points $(z_1, z_2, \cdots)$ with $|z_i| \leqq 1$ and let Z be the space gotten from Y by identifying any pair of points $\{z_i\}$, $\{z_i'\}$ which satisfy

* Norman Steenrod, Universal Homology Groups, *Amer. Jour. Math.* 58 (1936) pp. 661–701.

† C. H. Dowker, Hopf's Theorem for Non-Compact Spaces, *Proceedings of the National Academy of Sciences* 23 (1937), pp. 293–294.

$$|z_1| = |z_1'| = 1, \qquad \left(\frac{z_i}{z_i'}\right)^{2^i} = 1, \qquad i = 1, 2, \cdots .$$

By methods similar to those used above one can show that $H^2(Z, \mathfrak{J})$ is the limit of the direct system $\{G_i\}$, where G_i is a cyclic group of order 2^i and the homomorphisms h^{ik}, $i \leqq k$, are defined by the relation (3) of Example VIII 20.1. Consequently the group $H^2(Z, \mathfrak{J})$ is the group of dyadic rational numbers modulo 1, and its character group $H_2(Z, \mathfrak{R}_1)$ is the 0-dimensional dyadic group.

5. Mappings of compact spaces

In this section we discuss mappings of one compact space in another from the point of view of algebraic connectivity theory.

A) Let X and Y be two compact spaces and f a mapping of X in Y. Let G be a group and n an integer. Then f induces a homomorphism h^f of $H^n(Y, G) = H^n(Y)$ in $H^n(X, G) = H^n(X)$ as follows: Let σ be a covering of Y and σ_f the covering of X made up of the inverse-images of the members of σ. The vertices of $N(\sigma_f)$ are in 1:1 correspondence with the vertices of $N(\sigma)$ and this correspondence is obviously a simplicial mapping of $N(\sigma_f)$ in $N(\sigma)$. As in Proposition 3A) this simplicial mapping induces a homomorphism of $H^n(N(\sigma))$ in $H^n(N(\sigma_f))$. Moreover if τ is another covering of Y and $g_\sigma \varepsilon H^n(N(\sigma))$ and $g_\tau \varepsilon H^n(N(\tau))$ are related elements (see Definition VIII 12) so also are the images of g_σ and g_τ in $H^n(N(\sigma_f))$ and $H^n(N(\tau_f))$. Thus for each element of $H^n(Y)$ we have obtained an element of $H^n(X)$. It is this mapping of $H^n(Y)$ in $H^n(X)$ (easily seen to be a homomorphism) which we denote by h^f.

B) Similarly f induces a homomorphism h_f of $H_n(X)$ in $H_n(Y)$. As in Proposition 3 B) the simplicial mapping of $N(\sigma_f)$ in $N(\sigma)$ induces a homomorphism of $H_n(N(\sigma_f))$ in $H_n(N(\sigma))$. The desired homomorphism h_f of $H_n(X)$ in $H_n(Y)$ is obtained as follows: assign to each element e of $H_n(X)$ the element of $H_n(Y)$ whose representative in $H_n(N(\sigma))$ is the image of the representative of e in $H_n(N(\sigma_f))$.

Let f be a mapping of X in Y and g a mapping of Y in Z. Then the homomorphism of $H_n(X)$ in $H_n(Z)$ induced by gf is the product homomorphism $h_g h_f$. The analogous statement of transitivity can be made for the homomorphisms of cohomology groups induced by the mappings f, g, and gf.

Example VIII 23. Let P and Q be two polytopes in given simplicial decompositions. A *simplicial mapping* f of P in Q is one having the properties that if $p_0, \cdots, p_k$ are vertices of a cell of P then

$f(p_0), \cdots, f(p_k)$ are vertices of a cell of Q, and f maps the cell with vertices $p_0, \cdots, p_k$ *linearly* into the cell with vertices $f(p_0), \cdots, f(p_k)$. A simplicial mapping of P in Q induces, of course, a simplicial mapping of the vertex-scheme of P in the vertex-scheme of Q; and conversely a simplicial mapping of the vertex-scheme of P in the vertex-scheme of Q induces a simplicial mapping of P in Q. We leave to the reader the proof that if f is a simplicial mapping of P in Q the homomorphisms induced by f of $H_n(P)$ in $H_n(Q)$ and $H^n(Q)$ in $H^n(P)$ are the same as the homomorphisms, induced, as previously defined, by the simplicial mapping of the vertex schemes (see §3). (Consider the covering σ of stars of vertices in Q, the covering τ of stars of vertices in P, and use the fact that τ is a refinement of the covering formed by the inverse-images of the members of σ.)

Example VIII 24. Let τ be a covering of a space X, $P(\tau)$ the geometric realization of the nerve $N(\tau)$, and b a barycentric τ-mapping of X in $P(\tau)$ (see Definition V 9). Then the homomorphism h^b of $H^n(P(\tau))$ in $H^n(X)$ induced by b is simply the homomorphism assigning to each element e_τ of $H^n(N(\tau))$ the element e of $H^n(X)$ having e_τ as a representative. This follows immediately from the definition of h^b.

Definition VIII 15. Let C be a closed subset of a compact space X and f the identity mapping of C in X. The homomorphisms of $H_n(C)$ in $H_n(X)$ and $H^n(X)$ in $H^n(C)$ induced by f are called the *natural homomorphisms*. An element of $H_n(C)$ which is sent by the natural homomorphism into the zero of $H_n(X)$ is said to *bound* in X. An element of $H^n(C)$ which is the image under the natural homomorphism of an element of $H^n(X)$ is said to be *extendable* over X.

Remark. If X is a polytope and C a subpolytope then this definition is consistent with the definition (Definition VIII 7) of natural homomorphisms for the vertex-schemes of X and C, and this follows from Examples VIII 14 and VIII 23.

C) Let f be a mapping of a compact space X in a compact space Y and h^f the homomorphism of $H^n(Y)$ in $H^n(X)$ induced by f. Then for each element e of $H^n(Y)$ there is a positive number δ with the following property: If g is a mapping of X in Y such that $d(f, g) < \delta$ then $h^g(e) = h^f(e)$.

Proof. Given the element e of $H^n(Y)$, let $H^n(N(\sigma))$ be a group in which e has a representative, and let this representative be e_σ. Let the covering σ of Y be made up of the open sets $V_1, \cdots, V_k$ and let τ be a covering of X whose members $U_1, \cdots, U_k$ satisfy

$$\overline{U}_i \subset f^{-1}(V_i)$$

(the existence of τ follows from normality). Then the 1:1 correspondence between U_i and V_i yields a simplicial mapping m of $N(\tau)$ in $N(\sigma)$ which is the product of the simplicial mapping of $N(\sigma_f)$ in $N(\sigma)$ defined in A) and the projection mapping of $N(\tau)$ in $N(\sigma_f)$. m therefore induces a homomorphism of $H^n(N(\sigma))$ in $H^n(N(\tau))$, and it is clear that the image of e_σ under this homomorphism is an element of $H^n(N(\tau))$ representing $h^f(e)$. Put

$$\delta = \min d(f(\overline{U}_i), Y - V_i).$$

Then if g is a mapping of X in Y for which $d(f, g) < \delta$, $\overline{U}_i \subset g^{-1}(V_i)$, and hence g also induces the same simplicial mapping m of $N(\tau)$ in $N(\sigma)$. Consequently $h^g(e) = h^f(e)$.

D) Let f be a mapping of a compact space X in a compact space Y and h_f the homomorphism of $H_n(X)$ in $H_n(Y)$ induced by f. Then given any element e of $H_n(X)$ and a covering σ of Y there is a positive number δ with the following property: If g is a mapping of X in Y such that $d(f, g) < \delta$ then $h_f(e)$ and $h_g(e)$ have the same representative in $H_n(N(\sigma))$. This can be proved by means analogous to those used in C).

E) From C) and D) it follows that *homotopic mappings of one compact space in another induce the same homomorphism of cohomology groups, and the same homomorphism of homology groups.*

Example VIII 25. Let f be a mapping of a compact space X in the n-sphere S_n. Let $G = \mathfrak{J}$. By Example VIII 22 $H^n(S_n, \mathfrak{J})$ is a free cyclic group. We say that S_n is *oriented* when one of the two generators of this group has been singled out. Regarding S_n as oriented we define the *degree* of f, just as in Example VIII 16, as the element of $H^n(X, \mathfrak{J})$ corresponding under h^f to the generator of $H^n(S_n, \mathfrak{J})$. Also, again as in Example VIII 16, the degree of f can be specified as a character h_f of $H_n(X, \mathfrak{R}_1)$.

Proposition E) tells us that the degree of a mapping of a compact space in an oriented n-sphere is a *homotopy invariant.* We recall that we established a special case of this long ago, in §1 of Chapter IV, and there used it to demonstrate the non-contractibility of the n-sphere.

F) Let X and Y be compact spaces, G a countable group, G^* its character group, and n an integer. We recall (4 G)) that $H_n(X, G^*)$ and $H_n(Y, G^*)$ are the character groups of $H^n(X, G)$ and $H^n(Y, G)$. Now let f be a mapping of X in Y, h^f the homomorphism of $H^n(Y, G)$

in $H^n(X, G)$ induced by f, and h_f^* the homomorphism of $H_n(X, G^*)$ in $H_n(Y, G^*)$ induced by f. Then h_f^* is dual to h^f.

PROOF. Let σ be a covering of Y and σ_f the covering of X made up of the inverse images under f of the members of σ. Now $H_n(N(\sigma_f), G^*)$ and $H_n(N(\sigma), G^*)$ are the character groups of $H^n(N(\sigma_f), G)$ and $H^n(N(\sigma), G)$. Proposition F) is then an easy consequence of the observation that the homomorphism of $H_n(N(\sigma_f), G^*)$ in $H_n(N(\sigma), G^*)$ is dual to the homomorphism of $H^n(N(\sigma), G)$ in $H^n(N(\sigma_f), G)$ (see 3 C)).

6. Hopf's Extension Theorem

In this section the basis group used for cohomologies will always be the group $\mathfrak{J}$ of integers and the basis group for homologies will always be its character group, $\mathfrak{R}_1$.

Suppose C is a closed subset of a compact space X and f is a mapping of C in a compact space Y. Under what conditions is it possible to extend f over X (see Definition VI 2)? We have already studied this problem in Chapter VI using the methods of point-set topology. We shall obtain more precise information by investigating this problem anew, now from the point of view of algebraic topology.

We first state a necessary condition for the possibility of such an extension.

A) Let n be an integer and let h^f be the homomorphism of $H^n(Y)$ in $H^n(C)$ induced by f. In order that f be extendable over X it is necessary that every element of $H^n(C)$ which is an image of an element of $H^n(Y)$ under h^f be extendable over X (see Definition VIII 15).

PROOF. Suppose f could be extended over X, to a mapping F of X in Y. Let h be the natural homomorphism of $H^n(X)$ in $H^n(C)$; we recall that h is the homomorphism induced by the identity mapping of C in X. Now f is the result of the identity mapping of C in X followed by the mapping F; the formula $h^f = hh^F$ then shows that each element of $H^n(C)$ which is an image under h^f is also an image under h.

B) Observe that the necessary condition of Proposition A) is equivalent to the following homology condition: Each element of $H_n(C)$ which bounds in X is sent into zero by h_f.

PROOF. Let us denote by h_f the homomorphism of $H_n(C)$ in $H_n(Y)$ induced by f, and by h^* the natural homomorphism of $H_n(C)$ in $H_n(X)$. We recall that h_f and h^* are the duals of h^f and h. Now $h^fH^n(Y)$ and $hH^n(X)$ are subgroups of $H^n(C)$. The condition of Proposition A) requires that the first of these groups be contained in the second. This is equivalent to the statement that the annihilator of the first group should contain the annihilator of the second. But by Proposition 2D) the annihilator of $h^f H^n(Y)$ is the kernel of the homomorphism h_f, i.e. the set of elements of $H_n(C)$ which are sent into zero by h_f. Similarly the annihilator of $hH^n(X)$ is the kernel of the homomorphism h^*, i.e. the set of elements of $H_n(C)$ which bound in X. This shows the equivalence of the conditions of A) and B).

In general, the condition of A) is not sufficient.

EXAMPLE VIII 26. Let Y be a set in the plane consisting of two circumferences C_1 and C_2 tangent at a point P. Let X be a closed circular disc and C its boundary. Let f be the mapping of C in Y determined as follows. Divide C into four quadrants by the successive points P_1, P_2, P_3, P_4. Each of P_1, P_2, P_3, P_4 is mapped by f into P. The open arc P_1, P_2 is mapped topologically by f on the positively oriented arc $C_1 - p$, the open arc P_2, P_3 is mapped topologically by f on the positively oriented arc $C_2 - p$, the open arc P_3, P_4 is mapped topologically by f on the negatively oriented arc $C_1 - p$, and the open arc P_4, P_1 is mapped topologically by f on the negatively oriented arc $C_2 - p$. Then it can be shown that f is not homotopic to a constant, i.e. cannot be extended over X, despite the fact that h^f sends every 1-cocycle of Y into a cobounding cocycle.

EXAMPLE VIII 26.1. Let X be the closed region $x_1^2+x_2^2+x_3^2+x_4^2 \leqq 1$ in E_4 bounded by S_3. H. Hopf† has constructed a mapping of S_3 in S_2 which is not homotopic to a constant, i.e. cannot be extended over X with respect to S_2. Nevertheless, because $H^n(S_3) = 0$ for $n \leqq 2$ and $H^n(S_2) = 0$ for $n > 2$, every homomorphism of $H^n(S_2)$ in $H^n(S_3)$ is a zero homomorphism.

Hopf's Extension Theorem asserts that the condition of Proposition A) is not only necessary, but also sufficient, in case X has dimension $\leqq n + 1$ and Y is the n-sphere. This means that the topological problem of the extension of mappings is reduced in this case to a purely algebraic problem.

EXTENSION OF SIMPLICIAL MAPPINGS IN S_n

In preparation for the proof of Hopf's Extension Theorem we first consider simplicial mappings of polytopes and prove by simultaneous

† Über die Abbildungen der dreidimensionalen Sphäre auf die Kugelfläche, *Math. Ann.* 104 (1931), pp. 637–665.

induction the following two propositions (from now on n is an integer $\geqq 1$).

C_n) Let P be a polytope of dimension $\leqq n$ and R either an elementary n-sphere* R_n or one of its successive barycentric subdivisions. We regard R as oriented. Let f be a simplicial mapping of P in R. If the degree of f is 0, f is homotopic to a constant mapping.

D_n) Let P be a polytope of dimension $\leqq n + 1$, Q a subpolytope of P and f a simplicial mapping of Q in an oriented elementary n-sphere R_n. Let the element e of $H^n(Q)$ be the degree of f and the element $\tilde{e} \varepsilon H^n(P)$ an extension of e. Then f can be extended to a mapping F of P in R_n which has $\tilde{e}$ for its degree.

We first prove C_1), then we show that C_n) implies D_n), $n \geqq 1$, and finally we prove that D_n) implies C_{n+1}), $n \geqq 1$. This will establish C_n) and D_n) for all $n \geqq 1$.

PROOF OF C_1). Let $z_1^0 = (r_0, r_1)$ be a positively oriented 1-cell of the polygon R. Let φ be the 1-cocycle in P defined by

$$\begin{cases} \varphi(s_1) = \pm 1 \quad \text{if} \quad f(s_1) = \pm z_0, \\ \varphi(s_1) = 0 \text{ otherwise.} \end{cases}$$

Then φ represents the degree of f and by hypothesis φ is the coboundary of a 0-cocycle ψ. ψ is an integer-valued function of the vertices of the polygon P with this property: Let (p_0, p_1) be any 1-simplex of P. If $f(p_0, p_1)$ is the 1-simplex (r_0, r_1) then $\psi(p_1) - \psi(p_0) = 1$; if $f(p_0, p_1)$ is neither the 1-simplex (r_0, r_1) nor the 1-simplex (r_1, r_0) then $\psi(p_1) - \psi(p_0) = 0$.

We now identify the points of the polygon R with the elements of the group $\mathfrak{R}_1$ of reals mod 1. We may evidently assume that the segment z_1^0 directed from r_0 to r_1 corresponds to the interval $(0, \frac{1}{2})$. If p is a real number denote by (p) the congruence class mod 1 of p. We now define a real-valued function $F(x)$ over P by imposing on $F(x)$ two conditions, one a congruence and the other an inequality:

1° $(F(x)) = f(x)$,
2° if x is on the closed cell (p_0, p_1),

$$-\frac{1}{2} + \frac{\psi(p_0) + \psi(p_1)}{2} \leqq F(x) < \frac{\psi(p_0) + \psi(p_1)}{2} + \frac{1}{2} \cdot$$

* "Elementary sphere" is used in this section to denote both the complex defined in Example VIII 11 and its geometrical realization.

A simple computation shows that $F(x)$ is a single-valued continuous function. Defining, for $0 \leqq t \leqq 1$,

$$f(x, t) = (tF(x))$$

we see that f is homotopic to a constant mapping.

Proof that C_n) implies D_n). We denote by $r_0, r_1, \cdots, r_n, r_{n+1}$ the vertices of R_n in the order corresponding to the orientation and by z_n^0 the n-simplex $(r_1, r_2, \cdots, r_{n+1})$; z_n^0 is positively oriented. The cohomology class e is represented by the cocycle φ of Q defined by $\varphi(s_n) = 1$ if $f(s_n) = z_n^0$, $\varphi(s_n) = -1$ if $f(s) = -z_n^0$, and $\varphi(s_n) = 0$ otherwise, and the cohomology class $\tilde{e}$ is represented by a cocycle Φ of P which can be taken to be an extension of φ, by 1E).

Now let P_n and Q_n be the subpolytopes of P and Q made up of the cells of dimension $\leqq n$. Let $P_n^{(m)}$ be the mth successive barycentric subdivision of P_n modulo Q_n, m being large enough to permit the following construction: Denote by $s_n^1, s_n^2, \cdots, s_n^q$ all those n-simplexes of P_n which satisfy $\Phi(s_n^i) > 0$, $i = 1, \cdots, q$, and do not belong to Q. For each s_n^i we select $\Phi(s_n^i)$ n-simplexes in $P_n^{(m)}$:

$$s_n^{ik} = (p_1^{ik}, p_2^{ik}, \cdots \; p_{n+1}^{ik}), \qquad k = 1, \cdots, \Phi(s_n^i)$$

each s_n^{ik} being an oriented subsimplex of s_n^i having no common vertices with s_n^i nor with each other. We now extend f to a simplicial mapping F_1 of $P_n^{(m)} + Q$ in R_n by defining

$$F_1(p_l^{ik}) = r_l, \; i = 1, 2, \cdots, q; \; k = 1, 2, \cdots, \Phi(s_n^i); \quad l = 1, 2, \cdots, n+1,$$

$F_1(p) = r_0$ for the remaining vertices p of $P_n^{(m)} - Q$.

Aside from the simplexes of Q the only simplexes sent into z_n^0 by F_1 are the simplexes s_n^{ik}, and the degree of F_1 is represented by a chain Φ^* which is clearly related to Φ by formula (3) of p. 136 and consequently represents the same cohomology class of the polytope P regarded as a topological space. Hence we have

a) The degree of F_1 is the element of $H^n(P_n^{(m)}) = H^n(P_n)$ which corresponds to $\tilde{e}$ under the natural homomorphism of $H^n(P)$ in $H^n(P_n)$.

Now we have to show that F_1 can be extended over P. Let U be an $(n + 1)$-cell of $P - Q$. The mapping $F_1 | \overline{U} - U$ is a simplicial mapping of an n-dimensional polytope in R_n. Let $e^* \varepsilon H^n(\overline{U} - U)$ be the degree of $F_1 | \overline{U} - U$. Then the degree of F_1 is an extension of e^* over P_n and hence, by a), e^* is extendable even over P. A fortiori e^* is extendable over $\bar{U}$. But $H^n(\bar{U}) = 0$ (by Example VIII 10) and consequently $e^* = 0$. We now use C_n) to conclude that the mapping $F_1 | \bar{U} - U$ is homotopic to a constant mapping, or what is exactly the same $F_1 | \bar{U} - U$ is extendable over $\bar{U}$. It is clear that applying this reasoning to each $(n + 1)$-cell of $P - Q$ we obtain an extension F of F_1 over $\bar{U}$.

The degree of F is $\tilde{e}$, for the natural homomorphism of $H^n(P)$ in $H^n(P_n)$ is an isomorphism, and the image of the degree of F under this isomorphism is the degree of F_1, which by a) is also the image of $\tilde{e}$.

PROOF THAT D_n) IMPLIES C_{n+1}). P now stands for a polytope of dimension $\leqq n + 1$ and R either for an elementary $(n + 1)$-sphere or one of its successive barycentric subdivisions. Let z^0_{n+1} be a positively oriented $(n + 1)$-simplex of R and let V be the polytope spanned by those $(n + 1)$-simplexes of P which are sent into z^0_{n+1} by f. We may assume that no two of these simplexes have a common vertex, for otherwise we would replace P and R by their second barycentric subdivisions P'' and R'', and we would take for z^0_{n+1} an $(n + 1)$-simplex of R'' none of whose vertices is also a vertex of R (such a simplex obviously exists and satisfies our requirement). The homomorphism h^f of chains of R in chains of P sends the elementary chain $1z^0_{n+1}$ into an $(n + 1)$-chain φ of P which represents the degree of f, and hence by the hypothesis that the degree of f is zero, there exists an n-chain ψ such that

(1) $$\varphi = \gamma\psi.$$

Let $\tilde{\varphi}$ and $\tilde{\psi}$ denote the chains $h_V\varphi$ and $h_V\psi$ of V which correspond to φ and ψ under the natural homomorphism. From (1) we get

(2) $$\tilde{\varphi} = \gamma\tilde{\psi}.$$

Let z^0_n be an oriented face of z^0_{n+1} and let the n-chain ψ_1 of V be the image of the elementary chain $1z^0_n$ under the homomorphism induced by the mapping $f | V$. Since $1z^0_{n+1}$ is the coboundary of $1z^0_n$ in the complex spanned by z^0_{n+1} and $\tilde{\varphi}$ is the image of the chain $1z^0_{n+1}$ under the homomorphism induced by the mapping $f | V$, we have

$$\tilde{\varphi} = \gamma\psi_1,$$

and this together with (2) shows that the chain $\psi_1 - \tilde{\psi}$ of V is a cocycle in V and hence† (see Examples VIII 6 and VIII 10) cohomologous to 0 in V. Now let T be the polytope made up of the

† Here in order to apply Example VIII 6 we must take $n > 0$. This shows why we could start not our induction at $n = 0$.

simplexes of V of dimension $\leqq n$ and consider the mapping $f|T$ as a mapping of T in the elementary n-sphere R_n formed by the boundary simplexes of z_{n+1}^0. We may regard ψ_1 as a cocycle of T which represents the degree of $f|T$. Since $\psi_1 - \tilde{\psi}$, as shown before, is cobounding, the degree of $f|T$ is also represented by the chain $\tilde{\psi}$. Observe that by (1), $\gamma\psi(s_{n+1}) = 0$ for any $(n + 1)$-simplex which is not in V. This means that the chain $h_{(P-V)+T}\psi$ is an n-cocycle of the polytope $(P - V) + T$. Since this cocycle is an extension of $\tilde{\psi}$, we may use D_n) to conclude that $f|T$ can be extended to a mapping F of $(P - V) + T$ in R_n.

We set $g(x) = f(x)$ if $x \varepsilon V$ and $g(x) = F(x)$ otherwise; g is a mapping of P in R. Further we introduce in R a spherical metric such that the cell spanned by z_{n+1}^0 coincides with a hemisphere. It is clear that in this metric $f(x)$ and $g(x)$ are never antipodal and hence (Example VI 8) f and g are homotopic. Since the image of P under g is a proper part of the $(n + 1)$-sphere, namely the closed cell spanned by z_{n+1}^0, g is homotopic to a constant (see Example VI 8), and so is f.

This completes the proof that D_n) implies C_{n+1}), and consequently, by induction, establishes C_n) and D_n).

Mappings of compact spaces in S_n

Theorem VIII 1. *Let X be a compact space of dimension $\leqq n + 1$, C a closed subset of X, and f a mapping of C in an (oriented*) n-sphere S_n. Let $e \varepsilon H^n(C)$ be the degree of f. Then in order that f be extendable over X it is necessary and sufficient that e be extendable over X. Moreover, if $\tilde{e} \varepsilon H^n(X)$ is an extension of e over X there is an extension F of f over X having $\tilde{e}$ for its degree.*

Dually,

Theorem VIII 1′. Hopf's Extension Theorem. *Let X be a compact space of dimension $\leqq n + 1$, C a closed subset of X, and f a mapping of C in S_n. In order that f be extendable over X it is necessary and sufficient that every element† of $H_n(C)$ which bounds in X be sent into the zero of $H_n(S_n)$ by the homomorphism h_f of $H_n(C)$ in $H_n(S_n)$.*

The equivalence of Theorems VIII 1 and 1′ has been established in Proposition B). Thus it suffices to prove the first only.

* See Example VIII 25.

† We recall that we are using integers as basis for cohomology and reals mod 1 as basis for homology.

PROOF OF THEOREM VIII 1. The necessity of the condition has already been verified (Proposition A)).

To prove the sufficiency we first identify S_n with the elementary n-sphere R_n (see Proposition D_n)). We assert that there is a covering τ of X with these properties:

(a) $\tilde{e}$ has a representative, $\tilde{e}_\tau$, in $H^n(N(\tau))$,

(b) if $U_1, \cdots, U_k$ are the members of τ the image under f of U_iC is contained in the star of some vertex of R_n.

(c) the order of τ is $\leqq n + 1$.

To establish this we first find a covering τ_1 of X satisfying (a) and a covering τ_2 (whose existence follows from the compactness of X) satisfying (b). Then we take τ as any common refinement of τ_1 and τ_2 of order $\leqq n + 1$; the existence of this common refinement follows from Theorem V 1.

Denote by $\tau|C$ the covering of C made up of the intersections of the members of τ with C. dim $N(\tau) \leqq n + 1$ and $N(\tau|C)$ is a subcomplex of $N(\tau)$. Consider the *simplicial mapping* g_τ *of* $N(\tau|C)$ *in* R_n *obtained by assigning to each member* U_iC *of* $\tau|C$ *a vertex of* R_n *whose star contains* $f(U_iC)$. Now from the definitions of the homomorphisms of cohomology groups induced by mappings it follows easily that the *degree* e_τ *of* g_τ is a representative of e in $H^n(N(\tau|C))$. Consider the geometric realizations $P(\tau)$ and $P(\tau|C)$ of $N(\tau)$ and $N(\tau|C)$, and regard g_τ as a simplicial mapping of the polytope $P(\tau|C)$, and not just its vertices, in R_n. We may assume* that $\tilde{e}_\tau$ is an extension of e_τ. By D_n) we know that there is an extension G_τ of g_τ over $P(\tau)$ having $\tilde{e}_\tau$ as its degree. Now take a barycentric τ-mapping, b_τ, of X in $P(\tau)$ (see Definition V 9). The partial mapping $b_\tau|C$ is clearly a mapping of C in $P(\tau|C)$. Consider the mapping

$$g(x) = g_\tau(b_\tau(x)), \qquad x \in C,$$

of C in R_n and its extension

$$G(x) = G_\tau(b_\tau(x)), \qquad x \in X,$$

which is a mapping of X in R_n. The degree of g is the image of e_τ in $H^n(X)$ under the homomorphism induced by b_τ, and by Example VIII 24, this is the element e. By the same argument the degree of G is $\tilde{e}$.

The mapping g has the property that if $U_{i_0}, \cdots, U_{i_m}$ are all the members of τ containing a given point x of C then $g(x)$ is contained in

* If e'_τ is the element of $H^n(N(\tau|C))$ whose extension is $\tilde{e}_\tau$, e'_τ and e_τ represent the same element of $H^n(C)$; this means they have the same projections in $H^n(N(\tau_1|C))$, where τ_1 is a suitable refinement of τ. Replacing τ by τ_1, and g_τ by the product of g_τ and a projection of $N(\tau_1|C)$ in $N(\tau|C)$, we may assume that $e'_\tau = e_\tau$.

the cell of R_n spanned by the vertices $p_{i_0}, \cdots, p_{i_m}$ corresponding under g_τ to $U_{i_0}, \cdots, U_{i_m}$. The definition of g_τ implies that $f(x)$ is contained in the star of each vertex $p_{i_0}, \cdots, p_{i_m}$, i.e. $f(x)$ is contained in a cell of R_n having $(p_{i_0}, \cdots, p_{i_m})$ as a face; consequently $g(x)$ and $f(x)$ are contained in the same closed cell of R_n. Therefore f and g are homotopic, for we can join $g(x)$ to $f(x)$ by a straight-line segment and move $f(x)$ to $g(x)$ along this segment.

Now g admits an extension over X, namely G. Consequently (Borsuk's Theorem, Theorem VI 5) f also admits an extension F over X, which is homotopic to G. But G has the degree $\tilde{e}$; consequently (Proposition 5E)) F also has the degree $\tilde{e}$. This completes the proof of Theorem VIII 1.

Corollary 1. *If X is a compact space of dimension $\leqq n + 1$ to each element e of $H^n(X)$ corresponds a mapping of X in the oriented S_n with degree e.*

Corollary 2. *Let X be a compact space of dimension $\leqq n + 1$, and C a closed subset of X. In order that every mapping of C in S_n be extendable over X it is necessary and sufficient that every element of $H^n(C)$ be extendable, in other words that the natural homomorphism of $H^n(X)$ in $H^n(C)$ be a homomorphism of $H^n(X)$ on $H^n(C)$.*

Proof. Necessity: Suppose $e \in H^n(C)$ is not extendable. By Corollary 1 there is a mapping f of C in S_n with degree e. By Proposition A), f could not be extended over X. Sufficiency follows directly from Theorem VIII 1.

Corollary 3. *Let X be a compact space of dimension $\leqq n + 1$ and C a closed subset of X. In order that every mapping of C in S_n be extendable over X it is necessary and sufficient that only the zero element of $H_n(C)$ bound in X, in other words that the natural homomorphism of $H_n(C)$ in $H_n(X)$ be an isomorphism of $H_n(C)$ in $H_n(X)$.*

Proof. This is a consequence of Corollary 2 and Proposition 2F).

Theorem VIII 2. *Let X be a compact space* of dimension $\leqq n$. Two mappings of X in the oriented S_n are homotopic if and only if they have the same degree. Moreover, the homotopy classes of mappings of X in the oriented S_n are in 1:1 correspondence with the elements of $H^n(X)$.*

Proof. We have already proved that two homotopic mappings have the same degree (Proposition 5E)). We now prove the converse. Con-

* Observe that we now take dim $X \leqq n$, instead of dim $X \leqq n + 1$ as in Theorem VIII 1.

sider the topological product $X \times I$ of X and the interval I. By Theorem III 4, dim $X \times I \leqq n + 1$. Let p be the mapping of $X \times I$ in X defined by $p(x, t) = x$, $x \varepsilon X$, $0 \leqq t \leqq 1$, and q the mapping of X in $X \times I$ defined by $q(x) = (x, 0)$. Now p induces a homomorphism h^p of $H^n(X)$ in $H^n(X \times I)$, and q induces a homomorphism h^q of $H^n(X \times I)$ in $H^n(X)$. Since pq is the identity mapping of X on X the homomorphism $h^q h^p$ induced by this mapping is the identity automorphism of $H^n(X)$: for any element $e \varepsilon H^n(X)$

$$h^q h^p(e) = e. \tag{3}$$

Let X_0 be the set of points $(x, 0)$ and X_1 the set of points $(x, 1)$. If $n > 0$, $H^n(X_0 + X_1)$ is by Example VIII 22.1 the direct sum of $H^n(X_0)$ and $H^n(X_1)$. Hence the elements of $H^n(X_0 + X_1)$ can be represented by pairs (e, e'), e and e' being arbitrary elements of $H^n(X)$. From (3) it follows easily that given any element $e \varepsilon H^n(X)$ the element (e, e) of $H^n(X_0 + X_1)$ is the image of $h^p(e)$ under the natural homomorphism of $H^n(X \times I)$ in $H^n(X_0 + X_1)$, for the homomorphism h^q can obviously be interpreted as the natural homomorphism of $H^n(X \times I)$ in $H^n(X_0)$ (or in $H^n(X_1)$). This shows that every element of $H^n(X_0+X_1)$ of the type (e, e) is extendable over $X \times I$.

Let f_0 and f_1 be mappings of X in S_n having the same degree e. Consider the mapping of $X_0 + X_1$ defined as equal to f_0 on X_0 and f_1 on X_1. Its degree is (e, e) and (e, e) can be extended over $X \times I$. Hence by Hopf's Extension Theorem the mapping of $X_0 + X_1$ defined above can be extended over $X \times I$ also; this shows that f_0 and f_1 are homotopic.

The proof that homotopy classes are in 1:1 correspondence with cohomology classes now follows directly from Corollary 1 to Theorem VIII 1.

COROLLARY. *Let X be a compact space of dimension $\leqq n$. Then X admits an essential mapping in S_n if and only if $H^n(X) \neq 0$, or dually, if and only if $H_n(X) \neq 0$.*

REMARK. If X is a compact subset of Euclidean $(n + 1)$-space the hypothesis dim $X \leqq n$ in the Corollary can be dropped, for $H^n(X) \neq 0$ implies, by Corollary 1 to Theorem VIII 1, the existence of an essential mapping of X in S_n. On the other hand, from $H^n(X) = 0$ it follows (see Theorem VIII 1) that every mapping of X in S_n can be extended over an $(n + 1)$-cube containing X, and is consequently inessential (see Example VI 7). This allows us to give an algebraic interpretation to Theorem VI 13: *A compact subset X of E_{n+1} disconnects E_{n+1} if and only if $H^n(X) \neq 0$, or dually, if and only if $H_n(X) \neq 0$.*

This is, of course, a special case of Alexander's celebrated duality theorem.

7. Homology, cohomology, and dimension

We now come to the main result of Chapter VIII:

Theorem VIII 3. *Let X be a compact space of a finite dimension. In order that* dim $X \leqq n$ *it is necessary and sufficient that for any closed subset C of X every element of $H^n(C)$ be extendable over X. This means that the natural homomorphism of $H^n(X)$ in $H^n(C)$ must be a homomorphism of $H^n(X)$ on the entire group $H^n(C)$.*

Dually,

Theorem VIII 3′. *Let X be a compact space of a finite dimension. In order that* dim $X \leqq n$ *it is necessary and sufficient that given any closed subset C of X only the zero element of $H_n(C)$ bounds in X. This means that the natural homomorphism of $H_n(C)$ in $H_n(X)$ must be an isomorphism of $H_n(C)$ in $H_n(X)$.*

Since the natural homomorphism of $H_n(C)$ in $H_n(X)$ is the dual of the natural homomorphism of $H^n(X)$ in $H^n(C)$ (Proposition 5F) and Definition VIII 15), the conditions of both theorems are equivalent (Proposition 2F)). It suffices therefore to give only the

Proof of Theorem VIII 3. Necessity could be derived as an immediate consequence of Theorem VI 4 and Corollary 2 to Theorem VIII 1. A much more elementary proof, however, is the following: We first remark that if K is a complex of dimension $\leqq n$ and L a subcomplex of K every element of $H^n(L)$ is extendable over K, for there is no difference in K between n-chains and n-cocycles and every n-chain of L can obviously be extended to an n-chain of K.

Let e be an element of $H^n(C)$ and σ a covering of C such that e has a representative in $H^n(N(\sigma))$. Now consider a covering τ of X such that the members of σ are the intersections with C of the members of τ. By Theorem V 1 there is a refinement τ' of τ of order $\leqq n$. Let σ' be the covering of C made up of the intersection with C of the members of τ'. Since σ' is a refinement of σ, e has a representative, $e_{\sigma'}$, in $H^n(N(\sigma'))$. Because dim $N(\tau') \leqq n$ there is, as noted above, an element $\tilde{e}_{\tau'}$ of $H^n(N(\tau'))$ which is an extension of $e_{\sigma'}$. The element of $H^n(X)$ determined by $\tilde{e}_{\tau'}$ is an extension of e.

Sufficiency. Suppose dim $X > n$. Then by Proposition III 1 D), X contains a closed set X_1 of dimension $n + 1$. By Theorem VI 4 there exists a closed subset C of X_1 and a mapping f of C in S_n which cannot be extended over X_1. According to Theorem VIII 1 there is therefore, an element of $H^n(C)$ which cannot be extended over X_1, and a fortiori, cannot be extended over X.

REMARK: It is not known whether Theorem VIII 3 holds without the restriction to spaces of finite dimension.

RELATIVE HOMOLOGIES

In these last paragraphs we briefly discuss so-called "relative" homologies and cohomologies, by means of which the connections between dimension and homology could be expressed in neater form.

Given a compact space X and closed subset C we can define the homology and cohomology groups of X mod C. For each covering τ of X let $\tau | C$ be the covering of C made up of the intersections with C of members of τ. Then $\{H^n(N(\tau) \bmod N(\tau | C), G)\}$ (see page 116) is a direct system of groups with homomorphisms established by simplicial mappings of nerves. Similarly $\{H_n(N(\tau) \bmod N(\tau | C), G)\}$ is an inverse system of groups. The limit groups of these direct and inverse systems of groups are called the (n, G)-cohomology and (n, G)-homology groups of X mod C. The Propositions 1G) and 2H), proved for complexes, can be extended without difficulty to general compact metric spaces:

(a) If $H^{n+1}(X \bmod C, G)$ is zero then $H^n(X, G)$ is mapped on the entire group $H^n(C, G)$ by the natural homomorphism, i.e. every element of $H^n(C, G)$ is extendable.

(b) If G is a countable discrete group and G^* its character group then $H_n(X \bmod C, G^*)$ is the character group of $H^n(X \bmod C, G)$.

We now return to the use of the group $\mathfrak{J}$ for cohomologies and $\mathfrak{R}_1$ for homologies.

Theorem VIII 4. *Let X be a compact space of a finite dimension. In order that* dim $X \leqq n$ *it is necessary and sufficient that, given any closed subset C of X,* $H^{n+1}(X \bmod C) = 0$, *or equivalently* $H_{n+1}(X \bmod C) = 0$.

PROOF. The necessity follows by the same argument as that in Proposition 4F), and the sufficiency follows from Theorem VIII 3 and (a).

It can be shown that $H^m(X \bmod C)$ and $H_m(X \bmod C)$ are topological invariants of the open set $X - C$. Hence Theorem VIII 4 asserts that dim $X \geqq n$ if and only if X contains open sets which are carriers of essential n-dimensional homologies.

Appendix

Throughout this book the only spaces we have considered have been separable metric spaces, although spaces of more general nature have proved to be very important in recent topological investigations. It was not a matter of taste that decided us to consider only separable metric spaces, for a real theory of dimension applicable to general spaces would surely be very interesting.

Nor have we limited ourselves to separable metric spaces because there exist no definitions of dimension applicable to general spaces. Our reason is rather this: although it is possible—in fact, in several ways—to set up a *definition* of dimension for spaces of very general character, it is not possible (at least with any of our present concepts of dimension) to establish a *theory* of dimension for such general spaces. It is the purpose of the appendix to elucidate this contention.

Any theory of dimension starts with a "dimension function" $d(X)$, and it seems clear that in order to qualify for the name, a dimension function must be a numerical topological invariant of spaces, distinguishing between Euclidean n- and Euclidean m-spaces, and *monotone*, i.e. if $X' \subset X$ then $d(X') \leqq d(X)$.

In this book we have become familiar with the three dimension functions of separable metric spaces defined below ($d_1(X)$ is, of course, dim X):

(1) $d_1(X) = -1$ if X is empty; $d_1(X) \leqq n$ if for every point $p \in X$ and open set U containing p there is an open set V satisfying

$$p \in V \subset U,$$

$$d_1(\text{bdry } V) \leqq n - 1.$$

(2) $d_2(X) = -1$ if X is empty; $d_2(X) \leqq n$ if for every closed set $F \subset X$ and open set U containing F there is an open set V satisfying

$$F \subset V \subset U,$$

$$d_2(\text{bdry } V) \leqq n - 1.$$

(3) $d_3(X) = -1$ if X is empty; $d_3(X) \leqq n$ if every covering* of X has a refinement of order $\leqq n$.

We have proved (Proposition III 5A) and Theorem V 8) that, X being separable metric, the three statements $d_1(X) \leqq n$, $d_2(X) \leqq n$,

* We recall that "covering" means "finite covering by open sets."

$d_3(X) \leqq n$ are equivalent,* and by making use of these dimension functions a very attractive body of geometric theorems has been built up.

Now there is not the slightest difficulty in extending definitions (1), (2), or (3) to the most general sort of topological spaces. But one runs into trouble immediately thereafter, for, as will be shown by examples,

(*a*) d_1 does not fulfill the elementary requirement that if X is countable, $d_1(X) = 0$.

(*b*) $d_1(X)$ is not necessarily equal to either $d_2(X)$ or $d_3(X)$.†

(*c*) Neither d_2 nor d_3 is monotone.

d_1 is easily seen to be monotone, by the method used in Theorem III 1. However there is an example due to Urysohn‡ of a Hausdorff space U which, although countable, is connected. Now $d_1(U) > 0$; for a space X satisfying $d_1(X) = 0$ contains proper subsets which are both open and closed (in fact, arbitrarily small sets which are both open and closed) and hence X is disconnected. Urysohn's example therefore shows that the d_1 of a countable space need not be zero.

We shall now prove by making use of an example due to Tychonoff§ that neither d_2 nor d_3 is monotone. Tychonoff's example is constructed as follows: Let ω be the least ordinal of the second class and $[0, \omega]$ the space of all ordinals n, $0 \leqq n \leqq \omega$, the topology being the *order* topology, i.e. given any $n \varepsilon [0, \omega]$ a neighborhood of n is any set of x's of the form $\{m < x < p\}$ where $m < n < p$. Similarly let Ω be the least ordinal of the third class and $[0, \Omega]$ the space of all ordinals α, $0 \leqq \alpha \leqq \Omega$, the topology again being the order topology. Now consider the topological product S of $[0, \omega]$ by $[0, \Omega]$, so that the points of S are the pairs (n, α), $0 \leqq n \leqq \omega$, $0 \leqq \alpha \leqq \Omega$. Denote by T the complement in S of the single point (ω, Ω).

* It is appropriate to recall at this point the important role of this equivalence in demonstrating basic theorems of dimension theory: observe how we used d_2 to get the Sum Theorem for 0-dimensional Sets (Theorem II 2) and d_3 to get the Imbedding Theorem (Theorem V 3).

† See immediately preceding footnote.

‡ Über die Mächtigkeit zusammenhängender Mengen, *Math. Ann.*, 94 (1925), 262–295. The space U has a countable basis, but is not metric.

§ Über die topologische Erweiterung von Räumen, *Math. Ann.*, 102 (1930), 544–561.

Since these paragraphs were written the same example of Tychonoff has been employed, for very much the same purpose, by N. Vedenisoff: Remarques sur la dimension des espaces topologiques, *Uchenye Zapiski Moskov. Gos. Univ.*, 30 (1939), 131–140.

Each of $[0, \omega]$ and $[0, \Omega]$ is a compact* Hausdorff space. Hence† their product S is a compact Hausdorff space, and consequently‡ a normal space. On the other hand T is not normal: Let F be the set $\{(n, \Omega); 0 \leqq n < \omega\}$ and K the set $\{(\omega, \alpha); 0 \leqq \alpha < \Omega\}$. The sets F and K are disjoint, and closed in T; hence $U = T - K$ is an open set in T containing the closed set F. Now let V be any open set in T containing F; each point (n, Ω) of F then has a neighborhood contained in V: this means that for each n there is an ordinal $\alpha_n < \Omega$ such that $x > \alpha_n$ implies $(n, x) \varepsilon V$. But a countable collection of ordinals of the second class has an upper bound in the second class. Hence there is an ordinal $\alpha_0 < \Omega$ such that for each $n = 0, 1, 2, \cdots$ the point (n, α_0) is in V; therefore $\overline{V}$ must contain the point (ω, α_0). But (ω, α_0) is in K, proving that there exists no open set V satisfying $F \subset V \subset \overline{V} \subset U$; i.e. T is not normal.

Now it is easy to see that $d_1(S) = 0$. The compactness of S then shows (see the proof of Proposition II 4B)) that $d_2(S) = 0$. But $d_2(T) > 0$ since T is not normal, while a space X satisfying $d_2(X) = 0$ must be normal. To prove this last statement let F be a closed set in X and U an open set containing F. Because $d_2(X) = 0$ there is an open set V, $F \subset V \subset U$, such that the boundary of V is empty, i.e., V is also closed. Hence $F \subset V \subset \overline{V} \subset U$, proving that X is normal.

The results $d_2(S) = 0$ and $d_2(T) > 0$ show that d_2 is not monotone.

Let X be an arbitrary space. *Then* $d_2(X) = 0$ *implies* $d_3(X) = 0$. To show this we must prove that every covering α of X has a refinement made up of disjoint sets. First suppose α has only two members U_1 and U_2. In that case $X - U_1$ is a closed set contained in U_2. Because $d_2(X) = 0$ one can find a set V which is both open and closed and satisfies $X - U_1 \subset V \subset U_2$. Clearly $X - V$ and V make up the required refinement of α. Now suppose the number of members of α is $r > 2$; we then apply the method of Proposition V 1B), taking $M = X$ in that proposition.

Conversely, suppose that $d_3(X) = 0$; *then* $d_2(X) = 0$. For let F be a closed subset of X and U_1 an open set of X containing F. Then U_1 and $U_2 = X - F$ make up a covering α of X. The hypothesis $d_3(X) = 0$ tells us that α has a refinement β made up of disjoint open sets. Because the members of β are disjoint, each is closed as well as open; because β is a refinement of α, every member of β meeting F is

* In the sense of Bourbaki, = bicompact in the sense of Alexandroff-Urysohn.

† AH, p. 86.

‡ AH, p. 89.

contained in U_1. Let V denote the sum of the members of β meeting F. Then V is both open and closed and satisfies $F \subset V \subset U_1$; this proves that $d_2(X) = 0$.

Making use of the equivalence above between $d_2(X) = 0$ and $d_3(X) = 0$ we see that $d_3(S) = 0$ while $d_3(T) > 0$, so that d_3 also fails to be monotone. Thus in a theory of dimension based on either d_2 or d_3 it could happen, even for so "regular" a space as a bicompact Hausdorff space, that a subset had greater "dimension" then the entire space.

The relations $d_1(T) = 0$, $d_2(T) > 0$, $d_3(T) > 0$ prove assertion (b).

In this last paragraph we mention a very interesting unsolved problem raised by Menger:* that of giving an axiomatic characterization of dimension. Menger has solved this problem in the special case of subsets of the plane, showing that if $f(X)$ is a real-valued function defined for arbitrary subsets X of the plane which is

1. monotone: $X' \subset X$ implies $f(X') \leqq f(X)$,
2. F_σ-constant: X = countable sum of closed sets X_i implies $f(X) \leqq \text{Max}\, f(X_i)$,
3. topological: X homeomorphic to X' implies $f(X) = f(X')$,
4. compactifiable: every set X is homeomorphic to a subset of a compact set X' for which $f(X) = f(X')$,
5. normed: $f(\text{point}) = 0, f(\text{line}) = 1, f(\text{plane}) = 2$;

then $f(X) = \dim X$. The general problem appears to be quite difficult.

* Zur Begründung einer axiomatischen Theorie der Dimension, *Monatsh. f. Math. u. Phys.*, 36 (1929), pp. 193–218. See also Kuratowski and Menger, Remarques sur la théorie axiomatique de la dimension, *Monatsh. f. Math. u. Phys.*, 37 (1930), pp. 169–174, and Nöbeling, Die neuesten Ergebnisse der Dimensionstheorie, *Jahresbericht d. deutschen Mathem.-Vereinigung*, 41 (1932), pp. 1–16, in particular p. 15.

Index

The numbers refer to pages

Baire's Theorem, 56: *The countable intersection of open dense subsets of a complete space is dense.*

PROOF: Let $D_1, D_2, \cdots$ be open dense subsets of a complete space X, and D the intersection of the D_i. We have to show that every open set U of X meets D.

Because D_1 is dense there exists a point p_1 in UD_1, and because UD_1 is open it is possible to find a spherical neighborhood S_1 of p_1, of diameter less than 1, whose closure is contained in UD_1. If we now replace U by S_1 and D_1 by D_2 we get a new point p_2 and spherical neighborhood S_2 of p_2, of diameter less than $\frac{1}{2}$, whose closure is contained in S_1D_2. Continuing in this way we arrive at a sequence

$$(1) \qquad p_1, p_2, \cdots$$

such that

$$p_n \in S_n, \qquad \delta(S_n) < \frac{1}{n}, \tag{2}$$

and

$$\bar{S}_n \subset S_{n-1}D_n, \qquad \bar{S}_1 \subset UD_1. \tag{3}$$

Hence (1) is a Cauchy sequence and therefore has a limit p. It follows from (3) that $p \in UD$, which proves Baire's theorem.

Any open set is of course a G_δ-set. Hence Baire's theorem may be rephrased: *The countable intersection of dense G_δ-sets in a complete space is a dense G_δ.*

Brouwer's Reduction Theorem, 94: *In a space X with countable basis let $\{K_\lambda\}$ be a family of closed sets with this property: if*

$$K_1, K_2, K_3, \cdots$$

is a sequence of members of $\{K_\lambda\}$ such that

$$K_1 \supset K_2 \supset K_3 \supset \cdots,$$

then $K_1K_2K_3 \cdots$ is a member of $\{K_\lambda\}$. Then there exists an irreducible set in $\{K_\lambda\}$, i.e. a set $K \in \{K_\lambda\}$ which has no proper subset in $\{K_\lambda\}$.

Proof: Let

$$U_1, U_2, \cdots$$

be a countable basis of X, in some definite ordering. Let K_0 be an arbitrary member of $\{K_\lambda\}$. By induction we define sets K_n, $n = 1, 2, \cdots$, as follows:

$$K_n \in \{K_\lambda\} \tag{1}$$

and

$$K_n \subset K_{n-1} \cdot (X - U_n) \tag{2}$$

if such a set exists; otherwise $K_n = K_{n-1}$. Consider

$$K = \prod_1^\infty K_n. \tag{3}$$

Then K is irreducible; otherwise there would exist a proper subset K' of K belonging to $\{K_\lambda\}$, and therefore a U_n such that

$$KU_n \neq 0 \tag{4}$$

while $K'U_n = 0$. But then, by the construction of K_n, (2) would hold. However, this implies

$$KU_n = 0$$

by (3), contradicting (4).

List of Special Symbols

$B_n(K, G)$	Group of (n, G)-bounding cycles of the complex K, 112
$B^n(K, G)$	Group of (n, G)-cobounding cocycles of the complex K, 112
$\mathcal{C}$	Cantor set; dim $\mathcal{C} = 0$, 11
$C_n(K, G)$	Group of (n, G)-chains of the complex K, 110
$d(p, X)$	Distance between point p and set X
$d(x, y)$	Distance between points x and y
$d(X, Y)$	Distance between sets X and Y
E_n	Euclidean n-space, dim $E_n = n$, 41
E_ω	Hilbert space, 13
$f\|A$	If f is a mapping defined over a space X and A is a subset of X then $f\|A$ denotes the partial mapping in which f is considered as operating only on A, 75
F_σ-set	Countable sum of closed sets, 19
G_δ-set	Countable intersection of open sets, 56
h_K	Natural homomorphism of $H_n(L, G)$ in $H_n(K, G)$, 115
h_K	Natural homomorphism of $H^n(K \bmod L, G)$ in $H^n(K, G)$, 117
h_{K-L}	Natural homomorphism of $H_n(K, G)$ in $H_n(K \bmod L, G)$, 117
h_L	Natural homomorphism of $H^n(K, G)$ in $H^n(L, G)$, 115
$H_n(K, G)$	n-dimensional homology group of the complex K with basis group G, 112
$H^n(K, G)$	n-dimensional cohomology group of the complex K with basis group G, 112
$H_n(K \bmod L, G)$	n-dimensional homology group of the complex K modulo the subcomplex L, 116
$H^n(K \bmod L, G)$	n-dimensional cohomology group of the complex K modulo the subcomplex L, 116
$H_n(X, G)$	n-dimensional homology group of the compact space X with basis group G, 135
$H^n(X, G)$	n-dimensional cohomology group of the compact space X with basis group G, 135
I	Unit segment $0 \leqq x \leqq 1$ of the real line, 84
I_n	Cube in E_n, e.g. set of points each of whose n coordinates x_i satisfies $\|x_i\| \leqq 1$; dim $I_n = n$, 42

I_ω	Hilbert-cube, 12
$\mathcal{J}$	Set of irrational real numbers; dim $\mathcal{J} = 0$, 11
$\mathcal{J}_n$	Set of points in E_n all of whose coordinates are irrational; dim $\mathcal{J}_n = 0$, 11
$\mathcal{J}'_\omega$	Set of points in I_ω all of whose coordinates are irrational; dim $\mathcal{J}'_\omega = 0$, 13
$\mathfrak{J}$	Additive group of the integers, 108
K_n	Closed spherical region of E_n, e.g. the set of points in E_n whose distance from the origin is $\leqq 1$, 40
$\mathcal{L}_n^m$	Set of points in E_n at least m of whose coordinates are rational; dim $\mathcal{L}_n^m = n - m$, 29
$m_p(X)$	p-dimensional measure of set X, 103
$\mathcal{M}_n^m$	Set of points in E_n at most m of whose coordinates are rational; dim $\mathcal{M}_n^m = m$, 29
$\mathcal{M}_\omega^m$	Set of points in the Hilbert-cube at most m of whose coordinates are rational; dim $\mathcal{M}_\omega^m = m$, 29
$N(\alpha)$	Nerve of covering α, 68
$P(\alpha)$	Geometric realization of the nerve $N(\alpha)$, 68
$\mathcal{R}$	Set of rational real numbers; dim $\mathcal{R} = 0$, 11
$\mathcal{R}_n$	Set of points in E_n all of whose coordinates are rational; dim $\mathcal{R}_n = 0$, 11
$\mathcal{R}_n^m$	Set of points in E_n exactly m of whose coordinates are rational; dim $\mathcal{R}_n^m = 0$, 11
$\mathcal{R}_\omega$	Set of points in E_ω all of whose coordinates are rational; dim $\mathcal{R}_\omega = 1$, 13
$\mathcal{R}'_\omega$	Set of points in I_ω all of whose coordinates are rational; dim $\mathcal{R}'_\omega = 0$, 12
$\mathcal{R}_\omega^m$	Set of points in I_ω exactly m of whose coordinates are rational; dim $\mathcal{R}_\omega^m = 0$, 20
$\mathfrak{R}_1$	Additive group of the real numbers modulo 1, 108
s_n	n-simplex, 67
s_n	Oriented n-simplex, 109
S_n	n-sphere, e.g., set of points in E_{n+1} at distance 1 from origin, 37
$\mathcal{X}_n$	Nöbeling's universal n-dimensional space $= \mathcal{M}_{2n+1}^n \cdot I_{2n+1}$ = set of points in I_{2n+1} at most n of whose coordinates are rational, 64
Y^X	Functional space of mappings of X in Y, 55
$Z_n(K, G)$	Group of (n, G)-cycles of the complex K, 112
$Z^n(K, G)$	Group of (n, G)-cocycles of the complex K, 112
α-mapping	60

$\beta\varphi$	Boundary of the chain φ, 111
$\gamma\varphi$	Coboundary of the chain φ, 111
$\delta(X)$	Diameter of X
ϵ-mapping	57
π	Operator assigning a chain of K to each chain of the barycentric subdivision of K, 120
π'	Operator assigning to each chain of K a chain of the barycentric subdivision of K, 122
χ	Character of a group, 124
$\prec, \succ$	We write $s_n \prec s_{n+1}$ or $s_{n+1} \succ s_n$ if s_n is an oriented face of s_{n+1}, 109
$<\cdot$	Let K' be the barycentric subdivision of K; if $s_m' \varepsilon K'$ is an oriented subsimplex of $s_m \varepsilon K$, we write $s_m' <\cdot s_m$, 120